COSMIC CHEMICAL EVOLUTION

INTERNATIONAL ASTRONOMICAL UNION

UNION ASTRONOMIQUE INTERNATIONALE

COSMIC CHEMICAL EVOLUTION

PROCEEDINGS OF THE 187TH SYMPOSIUM OF THE INTERNATIONAL ASTRONOMICAL UNION, HELD AT KYOTO, JAPAN, 26–30 AUGUST 1997

EDITED BY

K. NOMOTO

School of Science,
Department of Astronomy & Research Center for the Early Universe,
Tokyo, Japan

and

K. W. TRURAN

University of Chicago & Enrico Fermi Institute,
Chicago, U.S.A.

KLUWER ACADEMIC PUBLISHERS

DORDRECHT / BOSTON / LONDON

A C.I.P. Catalogue record for this book is available from the Library of Congress.

ISBN 1-4020-0448-6

Published on behalf of
the International Astronomical Union
by
Kluwer Academic Publishers, P.O. Box 17, 3300 AA Dordrecht, The Netherlands.

Sold and distributed in North, Central and South America
by Kluwer Academic Publishers,
101 Philip Drive, Norwell, MA 02061, U.S.A.

In all other countries, sold and distributed
by Kluwer Academic Publishers,
P.O. Box 322, 3300 AH Dordrecht, The Netherlands.

Cover figure:
"Hubble's Deepest-Ever View of the Universe Unveils Myriad Galaxies
Back to the Beginning of Time
STScI-PRC96-01a – January 15, 1996, R. Williams and the HDF Team (ST ScI) and NASA"

Printed on acid-free paper

Printed in the Netherlands.

Contents

Preface xi

Conference Papers in order of Presentation:

Big Bang Nucleosynthesis 1
D.N. Schramm

Light Elements Abundances: New Insights on Stellar Mixing and Galactic Production 17
F. Primas

The First Stars 23
D. Arnett

Type 1A Supernovae and Chemical Evolution of Galaxies 33
K. Nomoto, C. Kobayashi, and H. Umeda

The Chemical Evolution of Light Elements in Our Galaxy 47
N. Prantzos

Nucleosynthesis in Red Giant Stars 57
N. Mowlavi

Fluorine Production from AGB and WR Stars 71
N. Mowlavi and G. Meynet

Deuterium Abundance in the Local ISM 75
J.L. Linsky and B.E. Wood

Abundances in Halo Population Stars 81
C. Sneden

Abundances in the Galactic Disk 91
B. Edvardsson

Abundances in a Hot Horizontal-Branch Star in the Old Open Cluster NGC/6791 97
R.C. Peterson and E.M. Green

On the Progenitors of Type Ia Supernovae 103
X.-D. Li and E.P.J. van den Heuvel

Abundances from z = 0 to z = 5 109
D.G. York

Enrichment of Heavy Elements in Clusters of Galaxies 117
Y. Ishimaru

On the Origin of Metallicity in Lyman-Alpha Forest Systems 123
M. Chiba and B.B. Nath

A Metal Enriched Dark Cluster of Galaxies at Z = 1 129
M. Hattori

Are Elliptical Galaxies Really Metal-Rich? 139
M. Loewenstein and R.F. Mushotzky

Chemical Evolution of Spiral Galaxies from Redshift 4 to the Present 147
U. Fritze- v. Alvensleben, U. Lindner, and K.J. Fricke

Age Problem Based on the Abundance Ratio from a High-Redshift QSO 153
T. Tsujimoto, Y. Yoshii, and K. Kawara

Stellar Yields and Chemical Evolution 159
B.K. Gibson

The Formation of First Generation Stars and Globular Clusters in Protogalactic Clouds 165
D.N.C. Lin and S.D. Murray

Constraints on Galaxy Formation from Extragalactic Globular Clusters 175
J.P. Brodie

The Evolution of Old Stellar Populations in Our Galaxy 185
S.R. Majewski

Chemical Evolution of the Magellanic Clouds 195
B. Barbuy, J.A. de Freitas Pacheco, and T. Idiart

Global Evolution of the Stars, Gas, Metals, and Dust in Galaxies 201
S.M. Fall

Conference Summary 209
B.E.J. Pagel

Conference Abstracts in Alphabetical Order

Nitrogen Abundance in Oxygen-Rich Giants through K to late M 219
W. Aoki and T. Tsuji

Asymmetrical Delayed Detonation From a 3D Hydrosimulation of a White Dwarf Explosion 220
E. Bravo ad D. García-Senz

Numerical Simulations of the Chemical Evolution of Galaxies Using Treecode-SPH 221
D.K. Churches, A.H. Nelson, and M.G. Edmunds

Chemical Evolution of Damped Lyman α Systems 222
K. Denda

Abundances in Metal-Rich Stars 223
S. Feltzing

Abundance Pattern of Supergiants in the Field of the Small Magellanic Cloud and in the Cluster NGC 330 224
V. Hill

Age and Metallicity Distribution of the Disk Stars from Edge-on Galaxies 225
A. Just

Evolution of the Colour-Magnitude Relation of Elliptical Galaxies 226
T. Kodama, N. Arimoto, A. Barger, and A. Aragón-Salamanca

The Optical Properties of Carbon and Graphite depend on Crystal Structure and Temperature 227
C. Koike, H. Shibai, T. Tanabe, H. Suto, H. Sogawa, C. Kaito, Y. Ando, and M. Fukada

Improved Theoretical Stellar Library for Population Synthesis Models 228
T. LeJeune and R. Buser

Chemical and Spectrophotometric Evolution of Optically Identified QSO Absorption Systems 229
U. Lindner, U. Fritze- v. Alvensleben, C.S. Möller, and K.J. Fricke

Comparison of Nucleosynthesis Models 230
O. Nakamura

The Galactic Disc Age-Metallicity Relation 231
Y.K. Ng, G. Bertelli, G. Carraro, and L. Portinari

Analysis of Elemental Carbon and Oxygen Abundances in Carbon Stars – They have turned out to be more Carbon-Rich than believed 232
K. Ohnaka and T. Tsuji

Abundances in Evolved Intermediate Mass Stars 233
M. Perinotto, L. Morbidelli, and P. Patriarchi

Spectral Energy Distributions of $z > 2$ Galaxies in the Hubble Deep Field 234
M. Sawicki and H.K.C. Yee

Blue 'Elliptical' Galaxies in the Hubble Deep Field 235
Y. Shioya and T. Yamada

The Symbiotic-Barium Stars Link 236
V.V. Smith, K. Cunha, A. Jorissen, and H.M.J. Boffin

The Spectral Evolution of Elliptical Galaxies in FIR and beyond 237
T. Takagi and N. Arimoto

Nucleosynthesis in ONEMG Novae: Comparison with Observations 238
S. Wanajo, K. Nomoto, and M. Hashimoto

Constraints on Cosmic Chemical Evolution from the Gamma-Ray Background 239
K. Watanabe, D.H. Hartmann, and M.D. Leising

Disk-Wind Models for Broad Absorption Line QSO's 240
T.M. Yamamoto

Supernova Observing Web (SNOW) Project with the Public Astronomical Observatories in Japan 241
H. Yamaoka, M. Doi, T. Shigeyama, N. Yasuda, and M. Watanabe

Element Abundances in Metal-Poor Stars 242
G. Zhao and Y.Q. Chen

Preface: Cosmic Chemical Evolution

Questions concerning the nature and origin of the chemical elements of which matter in the Universe is composed are as old as human inquiry. The history of the Universe itself is indeed written in the elemental and isotopic composition of its diverse components: galaxies, stars, interstellar gas in galaxies, the intergalactic medium, cosmic rays, and gas clouds at high redshifts. Since distinctive abundances and abundance patterns are characteristic of nucleosynthesis occurring in stars of different masses *and corresponding lifetimes*, interesting constraints on the star formation and nucleosynthesis histories may be contained in the abundance histories of both halo population stars and QSO absorption line systems. Using both distant quasars and most recently gamma-ray bursts, the star formation and abundance history of the high redshift Universe can also be probed. Significant progress has occurred in recent years as a consequence of a wealth of new information on cosmic phenomena - the spectroscopic and photometric properties of gas clouds and galaxies at high redshifts - pouring in from new ground and space based observatories, of improvements in our knowledge of the nuclear physics that dictates the character of explosive nucleosynthesis processes, and of rapidly expanding computational facilities which make possible detailed and more realistic modeling of stellar evolution and supernova nucleosynthesis.

The Universe emerged from an exciting first three minutes with a composition consisting of hydrogen (^{1}H), helium (^{4}He), deuterium (^{2}D), ^{3}He, and ^{7}Li. Further synthesis of heavy elements then awaited the formation and evolution of the first stars, some 10^8 to 10^9 years later. The characteristics of the first stellar contributions to nucleosynthesis, whether associated with Population II or a Population III, reflect (as might be expected) the nucleosynthesis products of the evolution of massive stars ($M \gtrsim 10M_\odot$) of short lifetimes ($\tau \lesssim 10^8$ years). Metal deficient stars ([Fe/H] $\lesssim$ -3) in our own galaxy's halo show two significant variations with respect to solar abundances: the elements in the mass range from oxygen to calcium are overabundant - relative to iron-peak nuclei - by a factor $\approx$ 2-3, and the abundance pattern in the heavy element regime A $\gtrsim$ 60-70 reflects the r-process abundance distribution that is characteristic of solar system matter, with no evidence for an s-process nucleosynthesis contribution. Both of these features are entirely consistent with nucleosynthesis expectations for massive stars and associated Type II supernovae. The introduction of both the s-process products of the evolution of low mass stars (M $\approx$ 1-3 $M_\odot$) and of iron nuclei from Type Ia supernovae occurs on a significantly longer timescale. In our Galaxy, evidence for low mass star input is evidenced at a metallicity -2.5 $\lesssim$ [Fe/H] $\lesssim$ -2, while the products of Type Ia supernovae seem first to have appeared at [Fe/H] $\sim$ -1. Such trends are consistent as well with abundance determinations for both galactic globular clusters and nearby dwarf galaxies. They are also consistent with observations of the abundances and of the metallicity evolution (in the high redshift Universe) of damped Lyman-α systems.

The considerations noted above have provided the motivation and the setting for IAU Symposium 187 on "Cosmic Chemical Evolution." The current level of activity and excitement is reflected in and captured by the diverse papers that have been collected into this volume. Indeed, symposium papers address virtually all aspects of the problem of nucleosynthesis, galactic chemical evolution, and cosmic chemical evolution. Theoretical models of nucleosynthesis in the Big Bang, in the first stellar generations, and in stars and supernovae over the histories of our Galaxy and other galaxies provide a measure of element production over the lifetime of the Universe. Observational studies of abundances in stars, galaxies, gas concentrations in galaxies and clusters of galaxies, the intergalactic medium, and gas clouds at high redshift are reviewed. To the extent that heavy element nucleosynthesis is dominated by stellar and supernova contributions, this abundance data provides a quantitative measure of the level of star formation activity as a function of time (redshift) that seems quite consistent with the starlight history of the Universe and reionization. Finally, studies of the chemical evolution of galaxies and expectations for the chemical evolution of the Universe itself provide a firm basis for comparison of theory and observation.

BIG BANG NUCLEOSYNTHESIS

D.N. SCHRAMM
The University of Chicago
5640 S. Ellis Avenue
Chicago, IL 60637, USA

Abstract. Big Bang Nucleosynthesis (BBN) is on the verge of undergoing a transformation now that extragalactic deuterium is being measured. Previously, the emphasis was on demonstrating the concordance of the Big Bang Nucleosynthesis model with the abundances of the light isotopes extrapolated back to their primordial values using stellar and Galactic evolution theories. Once the primordial deuterium abundance is converged upon, the nature of the field will shift to using the much more precise primordial D/H to constrain the more flexible stellar and Galactic evolution models (although the question of potential systematic error in ^{4}He abundance determinations remains open). The remarkable success of the theory to date in establishing the concordance has led to the very robust conclusion of BBN regarding the baryon density. The BBN constraints on the cosmological baryon density are reviewed and demonstrate that the bulk of the baryons are dark and also that the bulk of the matter in the universe is non-baryonic. Comparison of baryonic density arguments from Lyman-α clouds, x-ray gas in clusters, the Sunyaev-Zeldovich effect, and the microwave anisotropy are made and shown to be consistent with the BBN value.

1. Introduction

Big Bang Nucleosynthesis and the study of the light element abundances is undergoing a major transformation. The bottom line remains: primordial nucleosynthesis has joined the Hubble expansion and the microwave background radiation as one of the three pillars of Big Bang cosmology. Of the three, Big Bang Nucleosynthesis (BBN) probes the universe to far earlier times ($\sim$ 1 sec) than the other two and led to the interplay of cosmology with nuclear and particle physics. Furthermore, since the Hubble expansion

K. Nomoto and J.W. Truran (eds.), Cosmic Chemical Evolution, 1–15.

is also part of alternative cosmologies such as the steady state, it is BBN and the microwave background that really drive us to the conclusion that the early universe was hot and dense. The new extragalactic deuterium observations not only cement this picture and give added convergence on a value of the baryon density, Ω_b, they also enable BBN to become a constraint on stellar and Galactic evolution scenarios. It is this latter point that is the core of the transformation. Furthermore, new alternative methods of estimating the cosmic baryon density are now coming into use and are independently confirming the BBN prediction.

The current reviews will draw heavily on the recent reviews by Schramm (1997) in the *Proceedings of the National Academy of Sciences* and the proceedings of the Symposium on Light Element Synthesis held in Bern, Switzerland.

2. Overview

Although the extragalactic D/H observations have naturally attracted the most attention, it should not be forgotten that there are also recent heroic observations of ^{6}Li, Be and B, as well as ^{3}He and new ^{4}He determinations. Let us now briefly review the history, with special emphasis on the remarkable agreement of the observed light element abundances with the calculations. This agreement works only if the baryon density is well below the cosmological critical value. We will also note how a convergence on extragalctic D/H will enable powerful new constraints on stellar and Galactic evolution.

It should be noted that there is a symbiotic connection between BBN and the 3K background dating back to Gamow and his associates, Alpher and Herman. The initial BBN calculations of Gamow's group (Alpher *et al.*, 1948) assumed pure neutrons as an initial condition and thus were not particularly accurate, but their inaccuracies had little effect on the group's predictions for a background radiation.

Once Hayashi (1950) recognized the role of neutron-proton equilibration, the framework for BBN calculations themselves has not varied significantly. The work of Alpher, Follin and Herman (1953) and Tayler and Hoyle (1964), preceeding the discovery of the 3K background, and of Peebles (1966) and Wagoner, Fowler and Hoyle (1967), immediately following the discovery, and the more recent work of our group of collaborators (Copi *et al.*, 1997; Copi *et al.*, 1994; Walker *et al.*, 1991; Olive *et al.*, 1990; Schramm and Wagoner, 1977; Olive *et al.*, 1981; Yang *et al.*, 1984; Kawano *et al.*, 1988) all do essentially the same basic calculation, the results of which are shown in Figure 1.

As far as the calculation itself goes, solving the reaction network is rel-

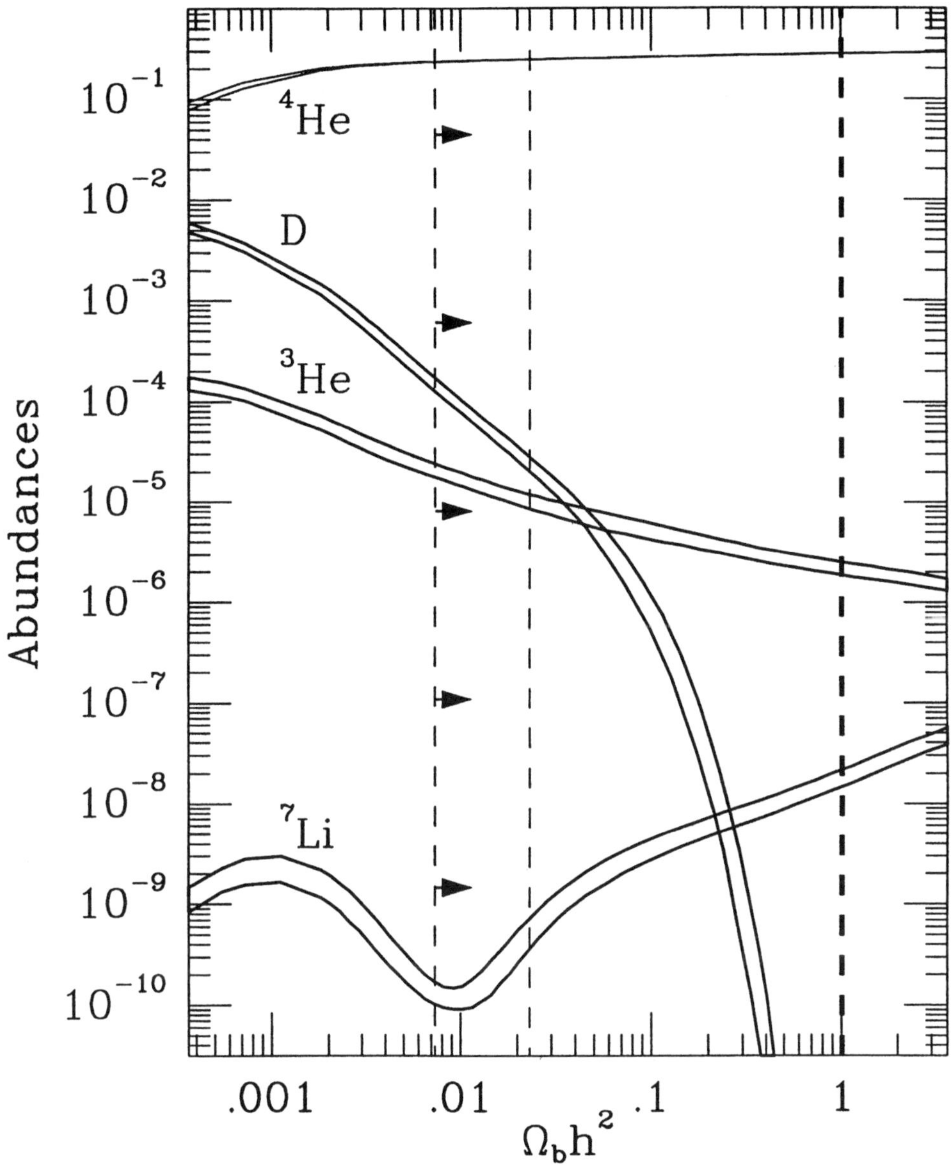

Figure 1. **Big Bang Nucleosynthesis abundance yields versus baryon density (Ω_b) and** $\eta \equiv \frac{n_b}{n_\gamma}$ for a homogeneous universe. ($h \equiv H_0/100$ km/sec/Mpc; thus, the concordant region of $\Omega_b h^2 \sim 0.015$ corresponds to $\Omega_b \sim 0.06$ for $H_0 = 50$ km/sec/Mpc.) Figure is from Copi, Schramm and Turner (1994). Note concordance region is slightly larger than Walker *et al.* (1991) due primarily to inclusion of possible systematic errors on Li/H. The width of the curves represents the uncertainty due to input of nuclear physics in the calculation. Recent measurements by Tytler, Fan and Burles (1996) narrow the vertical concordance region towards the high Ω_b side.

atively simple by the standards of explosive nucleosynthesis calculations in supernovae, with the changes over the last 25 years being mainly in terms of more recent nuclear reaction rates as input, not as any great calculational insight, although the current Kawano code (Kawano *et al.*, 1988) is somewhat streamlined relative to the earlier Wagoner code[6]. In fact, the earlier Wagoner code (Wagoner *et al.*, 1967) is, in some sense, a special adaptation of the larger nuclear network calculation developed by Truran (1965; Truran *et al.*, 1966) for work on explosive nucleosyntheis in supernovae. With the exception of Li yields and non-yields of Be and B (Steigman *et al.*, 1993), the reaction rate changes over the past 25 years have not had any major affect (see Yang *et al.*, 1984) and Krauss and his collaborators (Krauss and Romanelli, 1990; Kernan and Krauss, 1994), or Copi, Schramm, and Turner (1994) for a discussion of uncertainties). The one key improved input is a better neutron lifetime determination (Mampe *et al.*, 1989; Mampe *et al.*, 1993). There has been much improvement in the $t(\alpha,\gamma)$ ^{7}Li reaction rate, but as the width of the curves in Figure 1 shows, the ^{7}Li yields are still the poorest determined, both because of this reaction and even more because of the poorly measured ^{3}He (α,γ) ^{7}Be.

With the exception of the effects of elementary particle assumptions, to which we will return, the real excitement for BBN over the last 25 years has not really been in redoing the basic calculation. Instead, the true action is focused on understanding the evolution of the light element abundances and using that information to make powerful conclusions. In the 1960's, the main focus was on ^{4}He which is very insensitive to the baryon density. The agreement between BBN predictions and observations helped support the basic Big Bang model but gave no significant information, at that time, with regard to density. In fact, in the mid-1960's, the other light isotopes (which are, in principle, capable of giving density information) were generally assumed to have been made during the T-Tauri phase of stellar evolution (Fowler *et al.*, 1962), and so, were not then taken to have cosmological significance. It was during the 1970's that BBN fully developed as a tool for probing the universe. This possibility was in part stimulated by Ryter *et al.* (1970) who showed that the T-Tauri mechanism for light element synthesis failed. Furthermore, D abundance determinations improved significantly with solar wind measurements (Geiss and Reeves, 1971; Black, 1971) and the interstellar work from the Copernicus satellite (Rogerson and York, 1973). (Recent HST observations reported by Linsky *et al.* (1993) have compressed the local ISM D error bars considerably.) Reeves, Audouze, Fowler and Schramm (1973) argued for cosmological D and were able to place a constraint on the baryon density excluding a universe closed with baryons. Subsequently, the D arguments were cemented when Epstein, Lattimer and Schramm (1976) proved that no realistic astrophysical process other than the Big Bang could produce significant D. This baryon density was compared with dynamical determinations of density by Gott, Gunn, Schramm and Tinsley (1974). See Figure 2 for an updated H_0 - Ω diagram.

In the late 1970's, it appeared that a complimentary argument to D could be developed using ^{3}He. In particular, it was argued (Rood *et al.*, 1976) that, unlike D, ^{3}He was made in stars; thus, its abundance would increase with time. Unfortunately, recent data on ^{3}He in the interstallar medium (Gloeckler and Geiss, 1996) has shown that ^{3}He has been constant for the last 5 Gyr. Thus, low mass stars are not making a significant addition, contrary to these previous theoretical ideas. Furthermore, Rood, Bania and Wilson (1992) have shown that interstellar ^{3}He is quite variable in the Galaxy, contrary to expectations for a low mass star-dominated nucelus. However, the work on planetary nebulae shows that at least some low mass stars do produce ^{3}He. Nonetheless, the current observational situation clearly shows that arguments based on theoretical ideas about ^{3}He evolution should be avoided (c.f. Hata *et al.* (1995) where their "crisis" is really about ^{3}He problems [and excessively small assumed uncertainties in ^{4}He], not BBN]). Since ^{3}He now seems not to have a well understood history, simple ^{3}He or ^{3}He + D inventory arguments are misleading at best. However, onc is not free to go to arbitrary low baryon densities and high primordial D and ^{3}He, since processing of D and ^{3}He in massive stars also produces metals which are constrained (Copi *et al.*, 1995; Scully *et al.*, 1996) by the total metal content observed in the hot intra-cluster gas, if not the Galaxy. In the near future, this problem with ^{3}He evolution will be severely constrained by the extragalctic D/H. In particular, the Tytler and Burles $D/H = 3.4 \pm 0.3 \times 10^{-5}$ (Burles & Tytler 1997, Burles 1997) is somewhat higher than the pre-solar $D/H = 2.4 \pm 0.4 \times 10^{-5}$ (Geiss, 1997) and about a factor of 2 above the current interstellar $D/H = 1.5 \pm 0.1 \times 10^{-5}$ (Linsky, 1997). This tells us that the production of the current metal content of the Galaxy did not destroy more than 50% of the primordial D. Such low destruction could be accomplished with either Galactic infall or with a different stellar initial mass function at earlier times.

It was interesting that the abundances of the other light elements led to the requirement that ^{7}Li be near its minimum of $^7Li/H \sim 10^{-10}$, which was verified by the Pop II Li measurements of Spite and Spite (1982; Rebolo *et al.*, 1988; Hobbs and Pilachowski, 1988), hence yielding the situation emphasized by Yang *et al.* (1984) that the light element abundances are consistent over nine orders of magnitude with BBN, but only if the cosmological baryon density, Ω_b, is constrained to be around 6% of the critical value (for $H_0 \simeq 50$ km/sec/Mpc). The Li plateau argument was further strengthened with the observation of ^{6}Li in a Pop II star by Smith, Lambert and Nissen (1993). Since ^{6}Li is much more fragile than ^{7}Li, and yet it

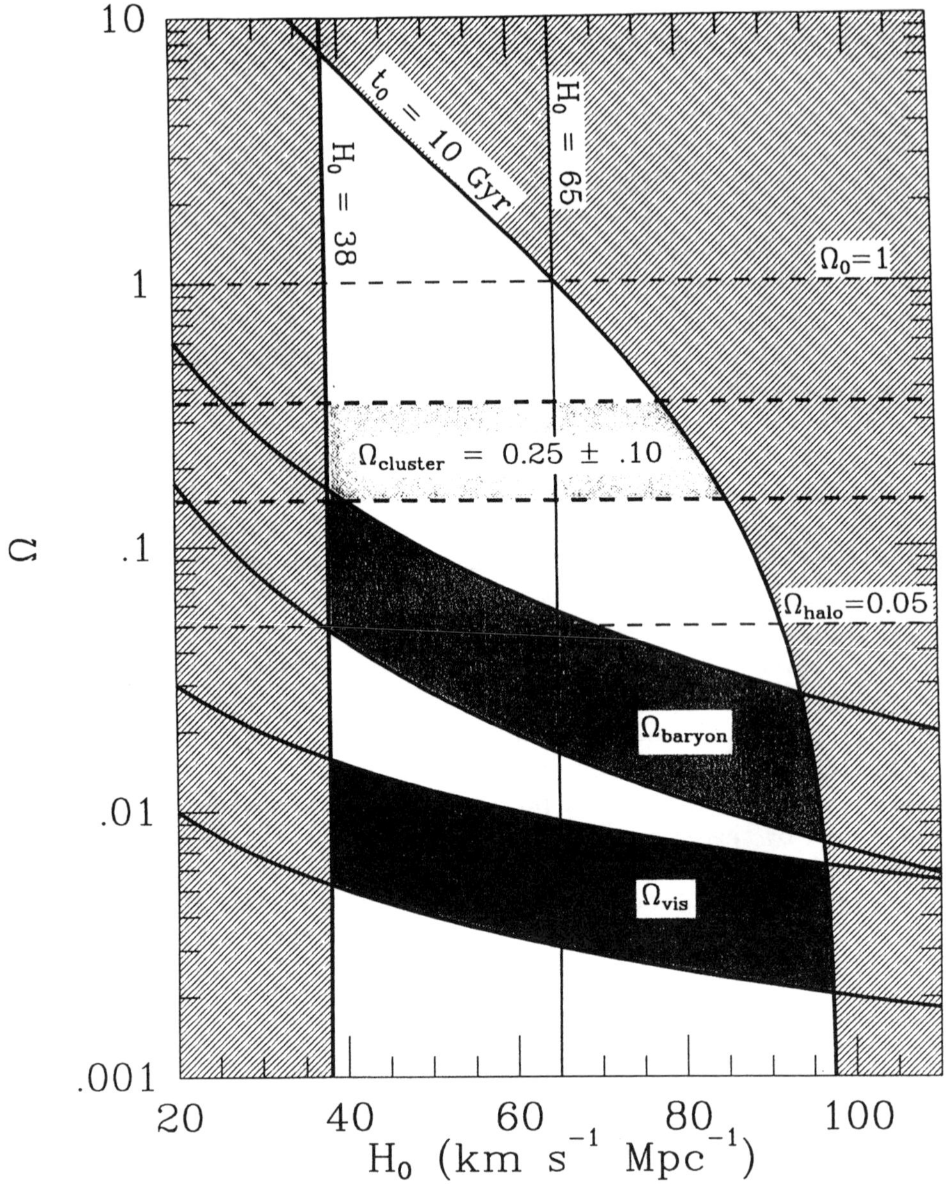

Figure 2. **An updated version of $H_0 - \Omega$ diagram of Gott, Gunn, Schramm and Tinsley (1974) showing that Ω_b does not intersect $\Omega_{VISIBLE}$ for any value of H_0 and that** $\Omega_{TOTAL} > 0.1$, so non-baryonic dark matter is also needed (Shi *et al.*, 1995).

survived, no significant nuclear depletion of ^{7}Li is possible (Steigman *et al.*, 1993; Olive and Schramm, 1992; Lemoine *et al.*, 1997). This observation of ^{6}Li was verified by Hobbs and Thorburn (1994). Lithium depletion mechanisms are also severly constrained by the recent work of Spite *et al.* (1996) showing that the lithium plateau is also found in Pop II tidally locked binaries. Thus, meridonal mixing is not causing significant lithium depletion. Recently Nollett *et al.* (1997) have discussed how ^{6}Li itself might eventually become another direct probe of BBN depending on the eventual low energy measurement of the D $(\alpha\gamma)$ ^{6}Li cross section and on spectroscopy improvements for extreme metal-poor dwarfs. With the new extragalactic D/H, one should be able to turn the ^{7}Li argument around and argue how much depletion and/or what model atmosphere is necessary. It is again clear from this argument that large amounts of depletion did not occur, contrary to the earlier models of Delyannis (1995).

Another development back in the 70's for BBN was the explicit calculation of Steigman, Schramm and Gunn (1977) showing that the number of neutrino generations, N_ν, had to be small to avoid overproduction of ^{4}He. (Earlier work (Tayler and Hoyle, 1963; Schvartzman, 1969; Peebles, 1971) had commented about a dependence on the energy density of exotic particles but had not done an explicit calculation probing N_ν.) To put this in perspective, one should remember that the mid-1970's also saw the discovery of charm, bottom and tau, so that it almost seemed as if each new detector produced new fundamental particle discoveries, and yet, cosmology was arguing against this "conventional" wisdom. Over the years, the limit on N_ν improved with ^{4}He abundance measurements, neutron lifetime measurements, and with limits on the lower bound to the baryon density, hovering at $N_\nu \lesssim 4$ for most of the 1980's and dropping to slightly lower than 4 just before LEP and SLC turned on (Walker *et al.*, 1991; Olive *et al.*, 1990; Schramm and Kawano, 1989; Pagels, 1991). This was verified by the results of the LEP Collaborations (1992; see also the CERN preprint CERN-PPE/96-183) where now the overall average is $N_\nu = 2.989 \pm 0.012$. A recent examination of the cosmological neutrino limit by Copi *et al.* (1997) in the light of the recent ^{3}He and D/H work shows that the BBN limit remains between 3 and 4 for all reasonable assumption options. It should be noted that this limit remains robust despite the uncertainties on ^{4}He systematics, since those uncertainties are still relatively small compared to a ΔN_ν of unity, although they are not small compared to significant shifts in ^{4}He implications for Ω_b.

The recent apparent convergence of the extra-galactic D/H measurements towards the lower values (Burles and Tytler, 1997; Hogan, 1998) D/H $\sim 3 \times 10^{-5}$ is beginning to collapse the Ω_B band in Figure 1 to a relatively narrow strip on the high Ω_B side (see arrows). However, such a full collapse at present is probably a bit premature. For example, Webb *et al.* (1997) find another high value of D and Cowie (1997) and his collegues still argue for possible high values. In any case, it is clear that deuteronomy (the study of deuterium in the cosmos) is a success since: 1) deuterium is clearly cosmological as it is seen in low metalicity and high redshift Lyman-α clouds; 2) the primordial D/H is higher than the present ISM D/H, as predicted by theory; and 3) the range of values for primordial D/H, regardless of whether or not the high or low ones win out, is consistent with the range of expectations based on the other light nuclei.

One potential problem that the low D/H, high Ω_B solution raises is the fact that the central primordial ^{4}He mass fraction is ~ 0.23, rather than ~ 0.245, which the Tytler and Burles (Tytler *et al.*, 1996; Tytler, 1997; Burles and Tytler 1997) D/H value would prefer for concordance. However, as Copi *et al.* (1997) emphasize, systematic uncertainties in Y_p cannot rule out such an excursion. But clearly we have to look carefully at ^{4}He. The recent work of Thuan *et al.* (1996) on $y \sim 0.24$ shows how uncertain the present situation is, but the resolution remains to be found. How high Y_p can be and still be consistent with the He observations in extragalactic H-II regions is still quite debatable, although most agree that Y_p up to 0.25 is not impossible. Schramm and Turner (1998) show that a Bayesian analysis of the plausible upper limit on Y_p centers on 0.25.

The power of homogeneous BBN comes from the fact that essentially all of the physics input is well determined in the terrestrial laboratory. The appropriate temperature regimes, 0.1 to 1 MeV, are well explored in nuclear physics laboratories. Thus, what nuclei do under such conditions is not a matter of guesswork, but is precisely known. In fact, it is known for these temperatures far better than it is for the centers of stars like our Sun. The center of the Sun is only a little over 1 keV, thus, below the energy where nuclear reaction rates yield significant results in laboratory experiments, and only the long times and higher densities available in stars enable anything to take place.

3. Density of Baryons

The bottom line that emerges from the above discussion is that (Copi *et al.*, 1997)

$$0.01 \leq \Omega_B h_0^2 \leq 0.025$$

where $h_0 \equiv H_0/100$ km/sec/Mpc. If the Tytler arguments on D/H do indeed hold up, then this will compress towards the high side, say $\Omega_B h^2 \sim 0.02 \pm 0.005$. Let us now compare with other ways of estimating Ω_b.

Attempts to circumvent the conclusion of homogeneous BBN by invoking a first-order quark-hadron phase transition (Applegate*et al.*, 1988; Alcock *et al.*, 1987) have merely illustrated the robustness of the conclusions. Figure 3 illustrates this fact, showing that for an optimized first-order quark-hadron phase transition, the abundances are only fit for the same

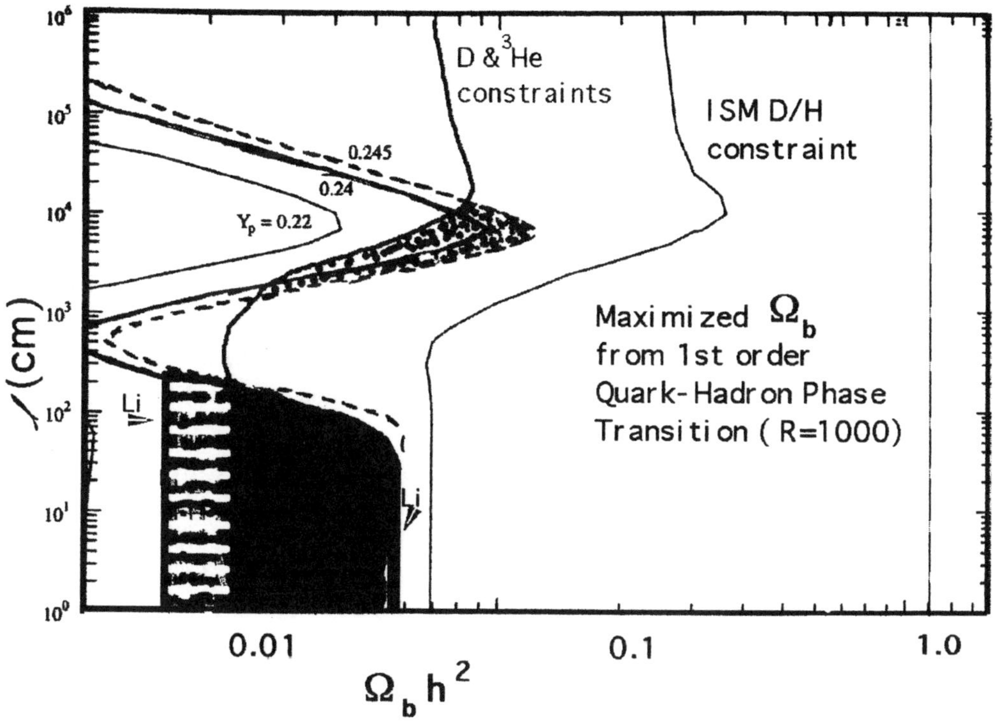

Figure 3. **An assumed first order quark-hadron phase transition still yields the same allowed range on $\Omega_b h^2$ when the same abundance constraints are used. The parameter** l is the separation of nucleation sites (the most sensitive additional parameter in the calculation). The dotted region is an additional allowed region if lithium is completely ignored. Even then, the constraint on Ω_b is within a factor of 2 of the standard value.

range in $\Omega_b h^2$ as in the homogeneous case. Only if the lithium constraint is completely ignored can higher $\Omega_b h^2$ values work, and even then, only a factor of 2 is possible.

3.1. LYMAN-α CLOUDS

Recent work by Bi and Davidsen (1997), by Quashnock and Vanden Berk (1997), and by Weinberg *et al.* (1997) also argues that the density of gas in the form of Lymon-α clouds at high redshift is consistent with the high end of the Big Bang Nucleosynthesis range on Ω_b. This would appear to resolve the long time problem of where are the "dark baryons." It is well known that $\Omega_{VISIBLE} \lesssim 0.01$, which, when compared to Ω_{BBN}, implies that the bulk of the baryons are not associated with stellar material. At least at high redshift this unseen material appears to have been found in these Lyman-α clouds. In conjunction with the Lyman-α clouds, it should also be noted that singly ionized helium is seen in the intergalactic gas, thus supporting the BBN fact that helium is primordial, and also supporting the point that significant numbers of baryons were between galaxies at high redshift (Jakobsen *et al.*, 1994; Bi and Davidsen, 1997).

3.2. HOT GAS IN CLUSTERS

Hot gas has been found in clusters of galaxies by ROSAT and ASCA. The temperature of the gas can be used to estimate the gravitational potential of the clusters if it is assumed that the gas is vivialized and purely supported by thermal pressure. Similarly, the intensity of the emission can be used to estimate the density of the gas. White *et al.* (1993) have shown that the typical values for x-ray clusters yield a hot gas to total mass ratio M_{HOT}/M_{TOT} of about 0.2.

Cluster masses can be estimated either from the temperature of the hot gas or from dynamics or from gravitational lensing. All yield the cluster implied density, $\Omega_{CLUSTER}$, of $\sim 0.25 \pm 0.10$. Thus, the implied bayon density from x-ray gas in clusters, $\Omega_{b,CLUSTER} \simeq \frac{M_{HOT}}{M_{TOT}} \times \Omega_{CLUSTER} \sim 0.05$, in good agreement with the BBN value for H = 50 and for the Tytler D/H.

3.3. SUNYAEV-ZELDOVICH EFFECT

Although not completed independent of the above x-ray gas determination, reliable maps of clusters are now being made using the Sunyaev-Zeldovich technique of looking for the distortion in the microwave background due to compton scattering from the hot, x-ray producing gas in the clusters (Carlstrom, 1998). The baryon density determined in this way has an h^{-1} dependence, whereas the pure x-ray gas determination goes as $h^{-3/2}$ (and BBN as h^{-2}). Carlstrom (private conversation, 1997) finds $\Omega_b \sim 0.06 \pm 02h^{-1}$ which is consistent with BBN for h$\sim$ 0.6.

3.4. MICROWAVE ANISOTROPIES

The method with the most potential for checking Ω_b is the measurement of the acoustic peaks in the microwave background anisotropy at angular scales near 1° or less (Jungman *et al.*, 1996a; 1996b). The height of the first doppler (acoustic) peak for gaussian fluctuation models is directly related to $\Omega_B h^2$, thus a direct check on BBN. Current experiments at the South Pole, at Sasketoon, and using balloons seem to favor values near the high side of the BBN range, but the present uncertainties are too large to make any strong statements. However, the next generation of satellites, NASA's MAP and ESA's PLANCK (formerly COBRAS/SAMBA) should be able to fix $\Omega_b h^2$ to better than 10% (if the sky is gaussian), which should provide a dramatic test of Big Bang Nucleosynthesis.

4. Dark Matter

The robustness of the basic BBN arguments and the new D/H measurements have given renewed confidence to the limits on the baryon density constraints. Let us convert this density regime into units of the critical cosmological density for the allowed range of Hubble expansion rates. This is shown in Figure 2. Figure 2 also shows the lower bound on the age of the universe of 10 Gyr from both nucleochronology and from globular cluster dating (Shi *et al.*, 1995) and a lower bound on H_0 of 38 from extreme type IA supernova models with pure 1.4 $M_\odot$ carbon white dwarfs being converted to ^{56}Fe. The constraint on Ω_b means that the universe *cannot be closed with baryonic matter.* (This point was made over twenty years ago (Reeves *et al.*, 1970) and has proven to be remarkably strong.) If the universe is truly at its critical density, then nonbaryonic matter is required. This argument has led to one of the major areas of research at the particle-cosmology interface, namely, the search for non-baryonic dark matter. In fact, from the lower bound on Ω_{TOTAL} from cluster dynamics of $\Omega_{TOTAL} > 0.15$, it is clear that non-baryonic dark matter is required unless $H_0 < 40$, regardless of whether $\Omega_{TOTAL} = 1$. Figure 2 also shows the range of $\Omega_{VISIBLE}$ and shows that there is no overlap between Ω_b and $\Omega_{VISIBLE}$. Hence, the bulk of the baryons are dark, that is, not in the form of stars.

Another interesting conclusion (Gott *et al.*, 1974) regarding the allowed range in baryon density is that it is in agreement with the density implied from the dynamics of single galaxies, *including their dark halos.* The recent MACHO (Alcock *et al.*, 1993) and (Aubourg *et al.*, 1993) reports of halo microlensing may well indicate that at least some of the dark baryons are in the form of baryonic objects in the halo. However, Gates, Gyuk and Turner (1995), and Alcock *et al.* (1997, 1993) show that the observed distribution of MACHOs favors less than 50% of the halo being in the form of MACHOS, but a 100% MACHO halo cannot be completely excluded, yet.

For dynamical estimates of Ω, one estimates the mass from $M \sim \frac{v^2 r}{G}$ where v is the relative velocity of the objects being studied, r is their separation distance, and G is Newton's constant. The proportionality constant out front depends on orientation, relative mass, etc. For large systems such as clusters, one uses averaged quantities. For single galaxies v would represent the rotational velocity and r, the radius of the star or gas cloud. It is this technique which yields the cluster bound on Ω shown in Figure 2. As mentioned above, the value of $\Omega_{CLUSTER} \sim 0.25$ is also obtained in those few cases where alignment produces giant gravitational-lens arcs (Fahlman *et al.*,1994) and also from the gravitational potential implied by the temperature of the observed x-ray gas in the clusters. As Davis (1997) showed, if the large scale velocity flows measured from the IRAS survey are due to gravity, then $\Omega_{IRAS} \gtrsim 0.2$. For $H_0 > 40$, $\Omega_{CLUSTER}$ already requires $\Omega_{TOTAL} > \Omega_{BARYON}$ and hence the need for non-baryonic dark matter.

An Ω of unity is, of course, preferred on theoretical grounds since that is the only long-lived natural value for Ω, and inflation (Guth, 1981; Linde, 1982) or something like it provided the early universe with the mechanism to achieve that value and thereby solve the flatness and smoothness problems. Note that our need for exotica is not dependent on the existence of dark galatic halos and that high values of H_0 increase the need for non-baryonic dark matter.

It is also interesting to note that the convergence of Ω on cluster scales at $\sim 0.25 \pm 0.1$ has important implications. If Ω_{TOTAL} is really unity, it would necessitate clusters not being a fair sample of the universe. Since standard CDM implies cluster scales as fair samples, this would imply a more complex structure formation picture. Options include biasing on cluster scales, a very hot dark matter component, or even a smooth background component such as a Λ_0 term, or a vacuum energy from a late-time phase transition (Hill and Schramm, 1985; Frieman *et al.*, 1996).

5. Acknowledgments

I would like to thank my collaborators, Craig Copi, Ken Nollett, Martin Lemoine, David Dearborn, Brian Fields, Dave Thomas, Gary Steigman, Brad Meyer, Keith Olive, Angela Olinto, Bob Rosner, Michael Turner, George Fuller, Sean Scully, Karsten Jedamzik, Rocky Kolb, Grant Mathews, Bob Rood, Jim Truran and Terry Walker for many useful discussions. I would further like to thank Art Davidsen, Poul Nissen, Jeff Linsky, David Tytler, Len Cowie, Scott Burles, Craig Hogan, Julie Thorburn, Doug Duncan, Lew Hobbs, Evan Skillman, Bernard Pagel and Don York for valuable discussion regarding the astronomical observations.

This work is supported by the NASA and the DoE (nuclear) at the University of Chicago, and by the DoE and by NASA grant NAG5-2788 at Fermilab.

References

Alcock, C. *et al.* (1993) *Nature* **Vol. no. 365**, 621-623.
Alcock, C. *et al.* (1997) *Astrophys. J.* **Vol. no. 486**, 697-726.
Alcock, C.R., Fuller, G. and Mathews, G. (1987) *Astrophys. J.* **Vol. no. 320**, 439-447.
Alpher, R.A., Bethe, H. and Gamow, G. (1948) *Phys. Rev.*, **Vol. no. 73**, 803-804.
Alpher, R.A., Follin, J.W. and Herman, R.C. (1953) *Phys. Rev.*, **Vol. no. 92**, 1347-1361.

Applegate, J.H., Hogan, C.J. and Scherrer, R.J. (1988) *Astrophys. J.* **Vol. no. 329**, 572-579.
Aubourg, E. *et al.* (1993) *Nature* **Vol. no. 365**, 623-625.
Bi, H.G. and Davidsen, A.F. (1997) *Astrophys. J.* **Vol. no. 479**, 523-542.
Black, D. (1971) *Nature* **Vol. no. 234**, 148-149.
Burles, S. (1997) Ph.D. Thesis, University of California at San Diego.
Burles, S. and Tytler, D. (1997) *Astrophys. J.*, submitted.
Carlstrom, J.E., Greco, L., Holzapfel, W.L. and Joy, M. (1998) in ed. A. Olinto, *Proc. of the 18th Texas Symposium on Relativisitic Astrophysics*, in press (World Scientific, Singapore).
Copi, C.J., Schramm, D.N. and Turner, M.S. (1994),*Science* **Vol. no. 267**, (192-(199.
Copi, C.J., Schramm, D.N. and Turner, M.S. (1995) *Astrophys. J.* **Vol. no. 455**, L95-L98.
Copi, C.J., Schramm, D.N. and Turner, M.S. (1997) *Phys. Rev. D* **Vol. no. 55**, 3389-3393.
Davis, M. (1997) in A. Olinto (ed.). *Proc. of the 18th Texas Symposium on Relativisitic Astrophysics*, in press (World Scientific, Singapore).
Delyannis, C.P. (1995) in *The Light Element Abundances* (Proc. of an ESO/EIPC Workshop held in Marciana Marina, Isola d'Elba, May 1994, ed. P. Crane, ESO Astrophysics Symposia (Springer-Verlag, Berlin), 395-409.
Epstein, R., Lattimer, J. and Schramm, D.N. (1976) *Nature* **Vol. no. 263**, 198-202.
Fahlman, G. Kaiser, N., Squires, G. and Woods, D. (1994) *Astrophys. J.* **Vol. no. 437**, 56-62.
Fowler, W.A., Greenstein, J. and Hoyle, F. (1962) *Geophys. J.R.A.S.* **Vol. no. 6**, 148-220.
Frieman, J.A., Hill, C.T., Stebbins, A. and Waga, I. (1996) *Phys. Rev. Lett.* **Vol. no. 75**, 2077-2080.
Gates, E., Gyuk, G. and Turner, M. (1995) *Phys. Rev. Lett.* **Vol. no. 74**, 3724-3727.
Geiss, J. (1997) in ed. C. de Jaeger, Proc. of International Space Science Institute Workshop on "Primordial Nuclei and Their Galactic Evolution," Bern, Switzerland, May (1997, *Space Sciences Series of ISSI*, in press (Kluwer Academic, Dordrecht).
Geiss, J. and Reeves, H. (1971) *Astron. and Astrophys.* **Vol. no. 18**, 126-132.
Gloeckler, G. and Geiss, J. (1996) *Nature* **Vol. no. 381**, 210-212.
Gott, J.R., III, Gunn, J., Schramm, D.N. and Tinsley, B.M. (1974) *Astrophys. J.* **Vol. no. (194**, 543-553.
Guth, A. (1981) *Phys. Rev. D* **Vol. no. 23**, 347-356.
Hata, N., Scherrer, R.J., Steigman, G., Thomas, D., Walker, T.P., Bludman, S. and Langacker, P. (1995) *Phys. Rev. Lett.* **Vol. no. 75**, 3977-3980.
Hayashi, C. (1950) *Prog. Theor. Phys.* **Vol. no. 5**, 224-235.
Hill, C.T. and Schramm, D.N. (1985) *Phys. Rev. D* **Vol. no. 31**, 564-580.
Hobbs, L. and Pilachowski, C. (1988) *Astrophys. J.* **326**, L23-L26.
Hobbs, L. and Thorburn, J. (1994) *Astrophys. J. Lett.* **428**, L25-L28.
Hogan, C. (1998) in ed. A. Olinto, *Proc. of the 18th Texas Symposium on Relativisitic Astrophysics*, in press (World Scientific, Singapore).
Jakobsen, P., Boksenberg, A., Deharveng, J.M., Greenfield, P., Jedrzewski, R. and Paresce, F. (1994) *Nature* **Vol. no. 370**, 35-39.
Jungman, G., Kosowsky, A., Kamionkowski, M., Spergel, D.N. (1996a) *Phys. Rev. Lett.* **Vol. no. 76**, 1007-1010.
Jungman, G., Kosowsky, A., Kamionkowski, M., Spergel, D.N. (1996b) *Phys. Rev. D* **Vol. no. 54**, 1332-1344.
Kawano, L., Schramm, D.N. and Steigman, G. (1988) *Astrophys. J.* **Vol. no. 327**, 750-754.
Kernan, P. and Krauss, L. (1994) *Phys. Rev. Lett.* **72**, 3309-3312.
Krauss, L.M. and Romanelli, P. (1990) *Astrophys. J.* **Vol. no. 358**, 47-59.
Kronberg, P.P. (1994) *Rep. Prog. Phys.* **Vol. no. 57**, 325-382.
Lemoine, M, Schramm, D.N., Truran, J.W. and Copi, C.J. (1997) *Astrophys. J.* **Vol. no. 478**, 554-562.

The LEP Collaborations: ALEPH, DELPHI, L3, and OPAL (1992) *Phys. Lett. B* **Vol. no. 276**, 247-253.
Linde, A. (1990) *Particle Physics and Inflationary Cosmology* (Harwood Academic Publishers, N.Y.).
Linsky, J. (1997) in Proc. of International Space Science Institute Workshop on "Primordial Nuclei and Their Galactic Evolution," Bern, Switzerland, May (1997, *Space Sciences Series of ISSI*, ed. C. de Jaeger, in press (Kluwer Academic, Dordrecht).
Linsky, J., Brown, A., Gayley, K., Diplas, A., Savage, B., Ayres, T., Landsman, W., Shore, S. and Heap, S. (1993) *Astrophys. J.* **Vol. no. 402**, 694-709.
Mampe, W. *et al.* (1993) *JETP Lett.* **Vol. no. 57**, 82-87.
Mampe, W., Ageron, P., Bates, C., Pendlebury, J.M. and Steyerl, A. (1989) *Phys. Rev. Lett.* **Vol. no. 63A**, 593-596.
Mushotsky, R. (1993) in eds. C.W. Akerlof and M.A. Srednicki, *Relativistic Astrophysics and Particle Cosmology Texas PASCOS 92, Annals of the N.Y Academy of Sciences* **Vol. no. 688**, 184-194.
Nollett, K., Lemoine, M. and Schramm, D.N. (1997) *Phys. Rev. C.* **Vol. no. 56**, 1144-1151.
Olive, K. and Schramm, D.N. (1992) *Nature* **Vol. no. 360**, 439-442.
Olive, K., Schramm, D.N., Steigman, G., Turner, M. and Yang, J. (1981) *Astrophys. J.* **Vol. no. 246**, 557-568.
Olive, K., Schramm, D.N., Steigman, G. and Walker, T. (1990) *Phys. Lett. B* **Vol. no. 236**, 454-460.
Pagel, B. (1991) *Physica Scripta*, **Vol. no. T36**, 7-15.
Peebles, P.J.E. (1966) *Phys. Rev. Lett.* **Vol. no. 16**, 410-413.
Peebles, P.J.E. (1971) *Physical Cosmology* (Princeton University Press).
Quashnock, J.M. and Vanden Berk, D.E. (1997) *Astrophyhs. J.*, submitted.
Rebolo, R., Molaro, P. and Beckman, J. (1988) *Astron. and Astrophys.* **Vol. no. (192**, (192-205.
Reeves, H., Audouze, J., Fowler, W.A. and Schramm, D.N. (1973) *Astrophys. J.* **Vol. no. 179**, 909-930.
Reeves, H., Fowler, W.A. and Hoyle, F. (1970) *Nature* **226**, 727-729.
Rogerson, J. and York, D. (1973) *Astrophys. J.* **Vol. no. 186**, L95-L98.
Rood, R.T., Bania, T., and Wilson, J. (1992) *Nature* **355**, 618-620.
Rood, R.T., Steigman, G. and Tinsley, B.M. (1976) *Astrophys. J.* **Vol. no. 207**, L57-
Ryter, C., Reeves, H., Gradstajn, E. and Audouze, J. (1970) *Astron. and Astrophys* **Vol. no. 8**, 389-397.
Schramm, D.N. (1997) 'Primordial Nucleosynthesis.' *PNAS*, **Vol. no. 94**, in press.
Schramm, D.N. and Kawano, L. (1989) *Nuc. Inst. and Methods A* **Vol. no. 284**, 84-88.
Schramm, D.N. and Wagoner, R.V. (1977) *Ann. Rev. of Nuc. Sci.* **Vol. no. 27**, 37-74.
Schvartzman, V.F. (1969) *JETP Letters* **Vol. no. 9**, 184-186.
Scott, D. and White, M. (1995) *Gen. Rel. and Grav.* **Vol. no. 27**, 1023-1030.
Scully, S.T., Cassé, M., Olive, K.A., Schramm, D.N., Truran, J. and Vangioni-Flam, E. (1996) *Astrophys. J.* **Vol. no. 462**, 960-968.
Shi, X., Schramm, D.N., Dearborn, D. and Truran, J.W. (1995) *Comments on Astrophys.* **Vol. no. 17**, 343-360.
Smith, V.V., Lambert, D.L. and Nissen, P.E. (1993) *Astrophys. J.* **Vol. no. 408**, 262-276.
Spite, M., Nissen, P.E. and Spite, F. (1996) *Astron. and Astrophys.* **Vol. no. 307**, 172-183.
Spite, J. and Spite, M. (1982) *Astron. and Astrophys.* **Vol. no. 115**, 357-366.
Steigman, G., Fields, B.D., Schramm, D.N., Olive, K. and Walker, T. (1993) *Astrophys. J.* **Vol. no. 415**, L35-L38.
Steigman, G., Schramm, D.N. and Gunn, J. (1977) *Phys. Lett. B* **Vol. no. 66**, 202-204.
Strickland, R. and Schramm (1997) *Astrophys. J.*, in press.
Tayler, R. and Hoyle, F. (1964) *Nature* **Vol. no. 203**, 1108-1110.
Thuan, T.X., Izotov, Y.I. and Lipovetsky, V.A. (1996) *Astrophys. J.* **Vol. no. 463**, 120-

133.
Truran, J.W. (1965) Doctoral Thesis, Yale University.
Truran, J. W., Cameron, A.G.W. and Gilbert, A. (1966) *Can. Jour. of Phys.* **Vol. no. 44**, 563-592.
Tytler, D. (1997) in eds. N. Prantos, M. Tosi and R. von Steiger, *Space Sciences Series of ISSI* (Proc. of ISSI Workshop on Primordial Nuclei and their Galactic Evolution, Bern, Switzerland, May 1997, in press (Kluwer, Dordrecht), and in *Space Science Reviews*, ed. C. de Jaeger, in press.
Tytler, D., Fan, X.-M. and Burles, S. (1996) *Nature*, **381**, 207.
Wagoner, R., Fowler, W.A. and Hoyle, F. (1967) *Astrophys. J.* **Vol. no. 148**, 3-49.
Walker, T., Steigman, G., Schramm, D.N., Olive, K. and Kang, H.S. (1991) *Astrophys. J.* **Vol. no. 376**, 51-69.
Webb, J.K., Carswell, R.F., Lanzetta, K.M., Ferlet, R., Lemoine, M., Vidal-Madjar, A. and Bowen, D.V. (1997) *Nature*, **Vol. no. 388**, 250-252.
Weinberg, D.H., Miralda-Escué, J., Hernquist, L. and Katz, N. (1997) *Astrophys. J.*, **Vol. no. 490**, 564-570.
White, S.D.M., Navarro, J.F., Evrard, A.E. and Frenck, C.S. (1993) *Nature* **Vol. no. 366**, 429-433.
Yang, J., Turner, M., Steigman, G., Schramm, D.N. and Olive, K. (1984) *Astrophys. J.* **Vol. no. 281**, 493-511.

LIGHT ELEMENTS ABUNDANCES: NEW INSIGHTS ON STELLAR MIXING AND GALACTIC PRODUCTION

F. PRIMAS
European Southern Observatory
Karl-Schwarzschild str. 2, D-85748 Garching b. München

1. Introduction

The remarkable finding that B follows Fe in almost direct proportion from the earliest times to the present, with little change of slope (if any) between halo and disk metallicities (cf Figure 1, open circles, Duncan et al. 1997), has stimulated the need for a revision of the classical cosmic-ray (CR) spallation scenario. A straightforward interpretation of Duncan et al. results is that the net rate of production of B (and Be too, since it shows a similar trend) does not depend on the CNO abundances in the ISM. It has been suggested that the CR spallation most important for light element production is not primarily protons and alpha particles colliding with CNO nuclei in the ISM (Reeves et al. 1970; Meneguzzi et al. 1971). Rather, it is C and O nuclei colliding with ambient protons and alpha particles, probably in regions of massive star formation (Cassé et al. 1995). This decouples light element production from the metallicity of the ISM and results in the approximately linear relationships observed, but some aspects are difficult to justify (see Prantzos, this volume).
Despite of the cosmological interest, lithium, beryllium, and boron are also sensitive probes of stellar interiors. They are destroyed in progressively deeper layers in the outer regions of stars by (p,α) reactions at about 2.5×10^6, 3.5×10^6, and 5.0×10^6 K respectively. Therefore, any detectable change in their abundances can be considered the observable consequence of internal processes. Only by gathering information on Li, Be, and B abundances in the same objects an overall understanding of light element production over Galactic history and stellar mixing can be successfully achieved. By combining a new set of B observations and the B data we had obtained in previous *Hubble Space Telescope* (HST) cycles, we are finally able to

K. Nomoto and J.W. Truran (eds.), Cosmic Chemical Evolution, 17–21.

gather a significant sample of stars for which Li, Be, and B abundances are known.

2. The Observed Sample

The boron 2500 Å spectral region has been observed with the Goddard High Resolution Spectrograph of the HST in a new set of stars with metallicities ranging between [Fe/H]=−2.0 and −1.0. With the choice of the G270M grating we obtain spectra of 26,000 resolution and typical S/N of 35 per pixel. We chose to target some Be-weak stars and few others characterized by very similar stellar parameters but with very different Li and/or Be abundances, with the aim of testing stellar mixing. The data has been reduced following the standard HST procedure using the IRAF **stsdas** package.

3. Abundance Analysis

The B abundances were determined via spectrum synthesis making use of the most recent Kurucz model atmospheres (Kurucz 1993a, 1993b). Slight adjustments to the line list tested by Duncan et al. (1997b) were introduced, in order to achieve a good overall match to the absorption features near the B I lines (cf Primas et al. 1997). The adopted solar B abundance is log n(B)=2.60 (Anders and Grevesse 1989).
A careful error analysis, combining uncertainties related to photon statistics in the points defining the line itself, continuum placement, and variations in the stellar parameters, was performed. In particular, the uncertainty in stellar metallicity is an indirect way of accounting for the likely existence of metal lines blending with the B feature. Since our spectra do not resolve the well-known blend between the Co I λ2496.716 feature and the B I λ 2496.772, then assuming a higher metallicity attributes more of the blend to the Co, decreasing the derived B abundance. We derived net errors ranging between 0.20 and 0.22 dex for each of our targets, and because of these negligible differences we decided to assign a conservative average error bar of 0.22 dex to each data point. A possible systematic error which has not been included yet could arise from NLTE effects. A more detailed description of our error analysis (including NLTE effects, which are under investigation) will be reported in Primas et al. (1997).

4. Discussion

Because of their low burning temperatures, circulation and destruction of the light elements can result in observable abundance changes, which in turn provide an invaluable probe of stellar structure and mixing. A large

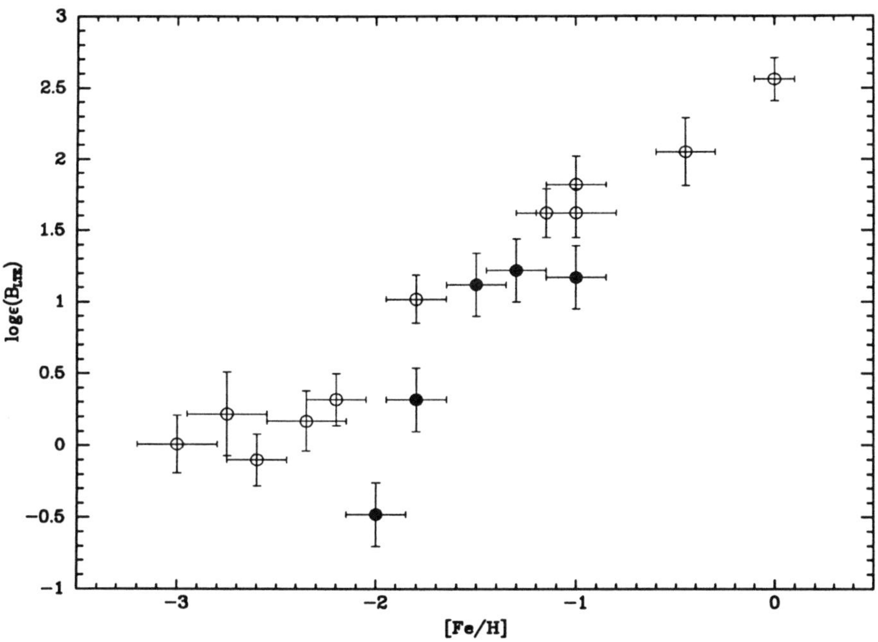

Figure 1. log $\epsilon(B_{LTE})$ vs. [Fe/H], combining the stars analyzed by Duncan et al. (1997a) and this work (filled circles)

amount of information can be obtained by studying the LiBeB pattern in single targets and/or by comparing spectra of stars with very similar stellar parameters, i.e. stars which have supposedly shared a common evolutionary history.

Although the most striking feature emerging from Figure 1 is the existence of real differences in the B content among stars at almost the same metallicity, for most of our stars, when analyzed singularly, the combined LiBeB information can be accounted for within known stellar structure scenarios. The only exception is HD 160617, a metal-poor ([Fe/H]≈ -1.80) subgiant star, with an effective temperature of about 6000 K. Its Li abundance is very close to the Spite plateau value (logN(Li)=2.2, where logN(Li)=log(N(Li)/N(H)) + 12), but it has a lower Be abundance with respect to stars with similar stellar parameters. Its subgiant status cannot account for any B depletion primarily because a 6000 K star is not expected to have experienced any dilution yet (Deliyannis et al. 1997). Even adopting the much lower effective temperature determined by Fuhrmann et al. (1994) for this star, i.e. T_{eff}=5664 K, does not justify the depletion we observe ($\approx$ a factor of 3). No standard nor non-standard stellar structure scenario account for a pattern in which Li is "unchanged" from what is currently considered its primordial value (the Spite plateau), while B (and Be) shows depletion. No firm conclusion can be drawn until different possible solutions will be further tested in the near future. The theoretical predictions of Be and B dilution factors might be incorrect: they have not been extensively tested, therefore light element abundances will need to be analyzed in a larger sample of subgiant stars, in order to better constrain the final numbers. HD 160617 might have formed with unusually low B (and possibly Be), which could explain its normal Li: we should be able to test such a picture by determining Be and B abundances in many more stars with similar stellar parameters, in order to get a sample statistically more significant. If we make the hypothesis that the primordial Li abundance was originally much higher, i.e. that the value logN(Li)=2.2 could reflect a depletion of a factor as large as $\approx$1.1 dex, it is then likely that the resulting LiBeB pattern might well resemble a standard light element depletion pattern.

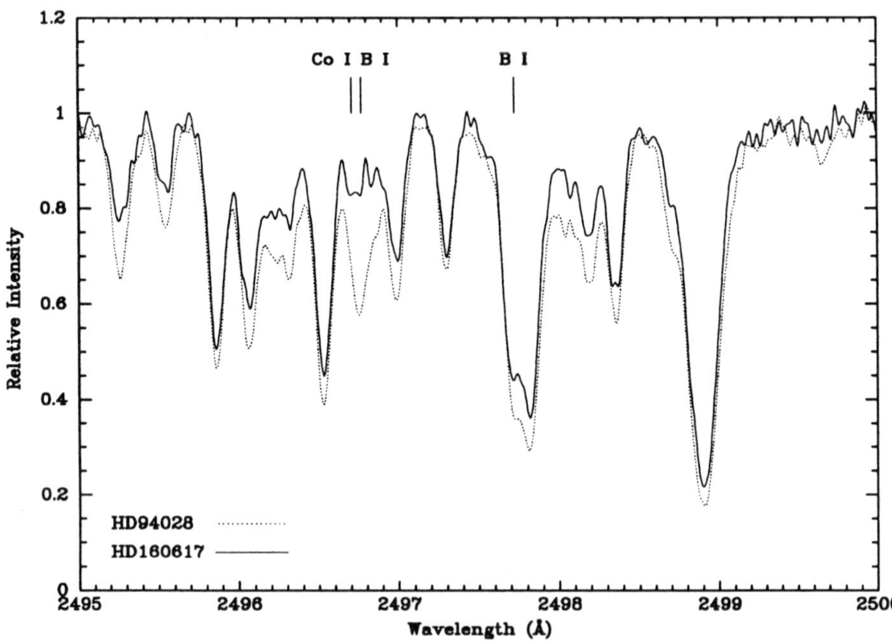

Figure 2. HD 160617 overplotted to HD 94028, a star with similar stellar parameters, although slightly more metal-rich

But a further piece of information can be obtained by inspecting Figure 2. When HD 160617 is compared to HD 94028, a star with similar stellar parameters, same Li abundance, but "normal" (i.e. *undepleted*) Be and B, it is clear that whatever causes the B (and Be) abundances to be different, does not grossly affect the Li content, as would be anticipated for light element depletion. A tentative conclusion could then be that the abundance contrasts observed in these two stars indicate inhomogeneity in their initial B (and Be) content rather than different light element depletion factors.

5. Conclusions

A new set of B observations has been analyzed with the aim of testing and better understanding depletion and/or mixing mechanisms in stellar interiors. We find real differences in the B content of stars characterized by very similar stellar parameters (T_{eff}, log g, [Fe/H]) that challenge the current predictions of different stellar depletion and mixing calculations. One case has been presented in more detail, and possible different interpretations of the results suggested. More data will need to be collected, in order to follow the steps we have pointed out. Only with a larger sample of data points available, it will be possible to address the new questions arisen from our analysis, e.g. if spatial variation of light element production in the Galaxy can or cannot be a valuable explanation for the observed dispersion.

References

Anders, E. and Grevesse, N.: 'Abundances of the Elements – Meteoritic and Solar', *Geochim. Cosmochim. Acta* **53**, pp. 197–214

Cassé, M., Lehoucq, R. and Vangioni-Flam, E.: 1995, 'Production and Evolution of Light Elements in Active Star-Forming Regions', *Nature* **373**, pp. 318–319

Deliyannis, C.P., Boesgaard, A.M., King, J.R. and Duncan, D.K.: 1997, 'New Observations of Beryllium in the Galactic Halo', *ApJ (submitted)*

Duncan, D.K., Primas, F., Rebull, L.M., Boesgaard, A.M., Deliyannis, C.P., Hobbs, L.M., King, J.R. and Ryan, S.G.: 1997a, 'The Evolution of Galactic Boron and the Production Site of the Light Elements', *ApJ* **488**, pp. 338–349

Duncan, D.K., Peterson, R.C., Thorburn, J.A. and Pinsoneault, M.H.: 1997b, 'Boron Abundances and Internal Mixing in Stars I: The Hyades Giants', *ApJ (submitted)*

Fuhrmann, K., Axer, M. and Gehren, T.: 1994, 'Balmer Lines in Cool Dwarf Stars II. Effective Temperatures and Calibration of Colour Indices', *A&A* **285**, pp. 585–594

Kurucz, R.L.: 1993a, 'ATLAS9 Stellar Atmospheres Programs and 2 km/s Grid', *Smithsonian Astroph. Obs.* **CD-ROM no. 13**

Kurucz, R.L.: 1993b, 'SYNTHE Spectrum Synthesis Programs and Line Data', *Smithsonian Astroph. Obs.* **CD-ROM no. 18**

Meneguzzi, M., Audouze, J. and Reeves, H.: 1971, 'The Production of the Elements Li, Be, B by Galactic Cosmic Rays in Space and its Relation with Stellar Observations', *A&A* **15**, pp. 337–359

Primas, F., Duncan, D.K. and Thorburn, J.A.: 1997, 'A New Set of B Observations: Implications for Stellar Mixing and Cosmology', *in preparation*

Reeves, H., Fowler, W.A. and Hoyle, F.: 1970, '', *Nature* **226**, pp. 727–729

THE FIRST STARS

DAVID ARNETT
Steward Observatory
University of Arizona
Tucson AZ 85721, USA

Abstract. The possible nature of the first generation of stars is considered, using a star of $25M_{\odot}$ as an example. General nucleosynthesis and the production of CNO catalysts is examined in detail. The increase in neutron excess and its significance for yields from explosive burning is discussed. An estimate of the ratio of ionizing photons to heavy elements produced is derived, for use in early universe simulations.

1. Introduction

The first stars are thought to produce the abundances seen in extreme population II objects, and the UV photons which re-ionize the universe (Gnedin & Ostriker, 1996). These stars are the culmination of an early epoch of structure formation, and may provide a new diagnostic tool for that era. They also bridge the gap between cosmological nucleosynthesis and the stellar nucleosynthesis currently seen.

2. Preliminary Considerations

2.1. INITIAL ABUNDANCES

The pattern of cosmological nucleosynthesis gives an interesting concordance (Wagoner, Fowler, & Hoyle, 1968; Schramm & Wagoner, 1979; Kolb & Turner, 1990) between predicted and inferred abundances of D, ^{3}He, ^{4}He, and ^{7}Li. Figure 1 shows this abundance pattern in more detail, as it appears upon re-calculation. The abundances are usually presented for those values above 10^{-12}, which is shown as a dotted line; here the lower limit is extended to 10^{-18}. The new nuclei are ^{7}Li, ^{9}Be, ^{10}B, ^{11}B, ^{12}C, ^{13}C, ^{14}C, and ^{14}N. All other nuclei are rarer. The total abundance of potential

K. Nomoto and J.W. Truran (eds.), Cosmic Chemical Evolution, 23–32.

catalysts for CNO burning is $X(CNO) \approx 10^{-14}$. As we shall see, these levels of abundance are inadequate for stable hydrogen burning.

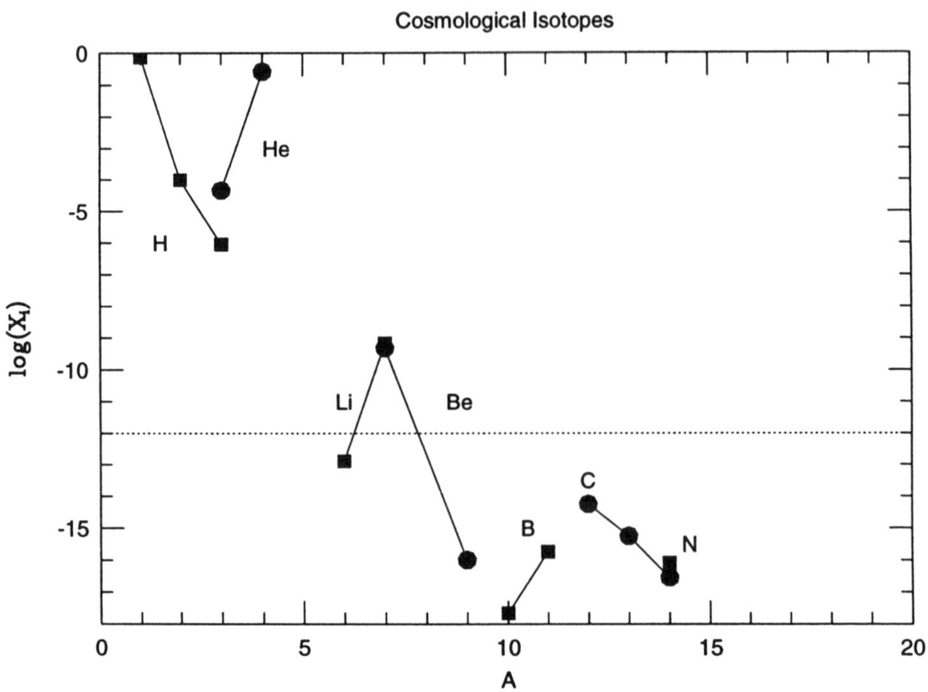

Figure 1. Cosmological Abundances.

2.2. REACTION NETWORK

While this recalculation of Big Bang nucleosynthesis used a network typical of previous work, a larger network was necessary for exploring subsequent nucleosynthesis. This network is shown in Figure 2; it contains 238 nuclei. The solid squares denote stable isotopes; the open squares are isotopes which are radioactive.

3. Failure of the pp-chain

An important feature of the evolution of the first stars is that, with the absence of CNO catalysts, the massive stars do not burn hydrogen under conditions of thermal balance. This may be simply shown. Following

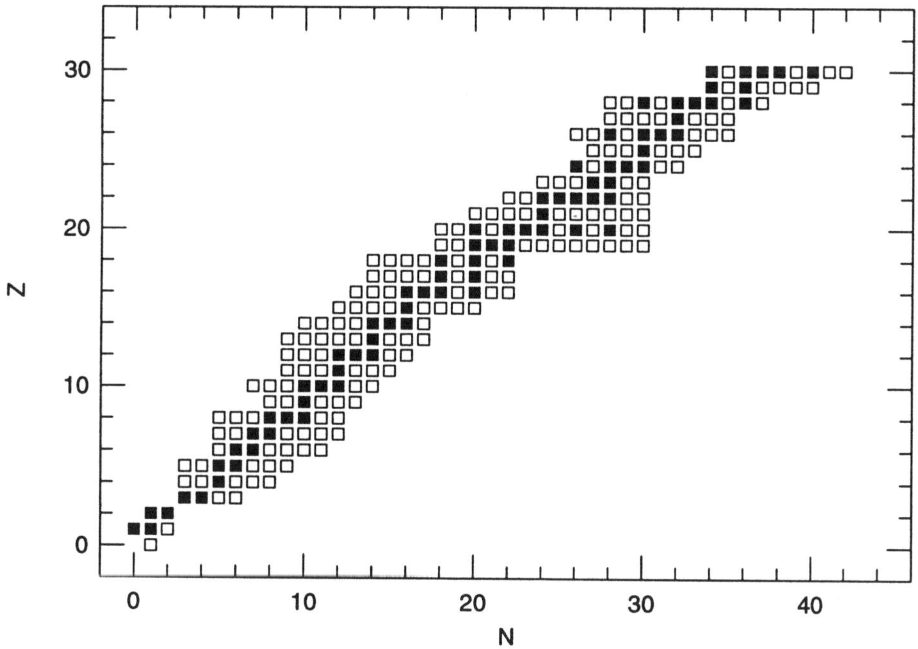

Figure 2. Nuclear Reaction Network.

Eddington, (see (Arnett, 1996) for detailed discussion),

$$L/M = 4\pi c G(1-\beta)/\kappa, \qquad (1)$$

where β, the ratio of gas to total pressure, may be related to radiation entropy per nucleon, $S_\gamma = \frac{4}{3}aT^3/N_A k\rho$, by

$$1-\beta = S_\gamma/(S_\gamma + 4Y), \qquad (2)$$

where Y is the total number of ideal gas particles per nucleon (ionized electrons plus ions).

For massive stars, which both produce ionizing radiation and synthesize new elements, the opacity κ is essentially just the Thomson scattering value, and $S_\gamma \approx 1$ for a star of 25 $M_\odot$. For a Salpeter initial mass function, such a mass corresponds to the typical star for bulk nucleosythesis.

For Big Bang abundances ($X_H = 0.77$, and $X_{He} = 0.23$), this gives

$$L/M = 9.3 \times 10^3 \text{ erg g}^{-1}\text{s}^{-1}. \qquad (3)$$

$T/10^8$ K	$\rho/(10^2$g/cc)	$\epsilon(pp)$ erg/g s	$\epsilon(CNO)$ erg/g s	$\epsilon(3\alpha)$ erg/g s
0.50	0.15	3.2E+02	7.6E-05	1.5E-09
0.57	0.23	6.6E+02	6.8E-04	9.2E-08
0.66	0.35	1.4E+03	5.5E-03	4.8E-06
0.76	0.53	2.7E+03	4.2E-02	2.1E-04
0.87	0.80	5.4E+03	2.9E-01	7.6E-03
1.00	1.20	1.0E+04	1.9E+00	2.4E-01
1.15	1.82	2.0E+04	1.1E+01	6.8E+00
1.32	2.76	3.8E+04	6.4E+01	1.9E+02
1.51	4.18	7.1E+04	3.4E+02	7.6E+03
1.73	6.32	1.3E+05	1.7E+03	3.4E+05
1.99	9.57	2.4E+05	8.1E+03	1.2E+07

TABLE 1. Energy generation rates for $M = 25M_\odot$ ($S_\gamma = 1$).

For thermal balance, in which the energy radiated is replaced by energy from nuclear burning, $<\epsilon>= L/M$. This gives a constraint on the energy release necessary for stable burning.

This may be compared with the actual energy generation rates from the pp-chain, the CNO cycle, and the triple-alpha process, which are given in Table 1. The rates are not drastically different from classical estimates (Clayton, 1968). We see that temperature of order 10^8 K are needed to attain thermal balance by the pp-chain, $<\epsilon(pp)>\approx 10^4$. This is shown in the third column. The fourth column gives the CNO energy generation rate, using the Big Bang value of $X(CNO) \approx 10^{-14}$. Despite its more sensitive dependence upon temperature, the CNO rate is smaller than the pp-chain rate throughout the temperature range shown. In the last column is given the energy generation rate for helium burning by the triple-alpha reaction, $\epsilon(3\alpha)$, which exceeds $\epsilon(pp)$ for temperatures $T > 1.7 \times 10^8$ K. The time scale for thermal cooling is $\tau_{thermal} = E/\epsilon \approx 1.5 \times 10^{12}$ s. These features have implications for the evolution of the star.

The time for consumption of hydrogen is $\tau_H = -X_H \,/\, (dX_H/dt)$, which is given in Table 2. The rate of energy generation by the CNO cycle is proportional to the abundance of catalyst nuclei available,

$$\epsilon_{CNO} = X(CNO)X(H)f(\rho, T). \tag{4}$$

For what catalyst abundance $X(CNO)$, does $\epsilon(CNO)$ balance the energy loss by radiation, $L/M \approx 10^4$ erg/g s? This is denoted by $X(CNO)$ in Table 2. For abundance of catalysts appropriate to the Big Bang, $X(CNO) \approx 10^{-14}$, the star would have to reach temperatures $T \approx 2 \times 10^8$ K to balance radiation loss by CNO burning.

$T/10^8$ K	$\rho/10^2$(g/cc)	τ_H(s)	$X(CNO)$	$\Delta X(CNO)$
0.50	0.15	1.7E+16	1.3E-06	1.3E-15
0.57	0.23	8.2E+15	1.5E-07	8.0E-14
0.66	0.35	4.0E+15	1.8E-08	4.1E-12
0.76	0.53	2.0E+15	2.4E-09	1.8E-10
0.87	0.80	1.0E+15	3.5E-10	6.6E-09
1.00	1.20	5.1E+14	5.3E-11	2.1E-07
1.15	1.82	2.7E+14	8.8E-12	6.2E-06
1.32	2.76	1.4E+14	1.6E-12	2.3E-04
1.51	4.18	7.6E+13	2.9E-13	1.4E-02
1.73	6.32	4.2E+13	5.8E-14	7.8E-01
1.99	9.57	2.3E+13	1.2E-14	3.0E+01

TABLE 2. **Burning Paramerters for $M = 25M_\odot$ $(S_\gamma = 1)$.**

Suppose the star contracts on a timescale $\tau_{thermal}$ to $T \approx 10^8$ K where $<\epsilon> \approx L/M$ During this time the triple alpha reaction would increase the abundance of CNO by

$$\Delta X(CNO) \approx \frac{dX(3\alpha)}{dt} \tau_{thermal}. \tag{5}$$

This crude estimate suggests that as the star approaches thermal balance by the pp-chain, the triple alpha process enhances the abundance of CNO catalysts by a sufficient amount to allow CNO cycle burning to take over.

4. Nucleosynthesis

4.1. TO THE END OF HYDROGEN BURNING

Figure 3 shows the detailed evolution of abundances using the network of Figure 2. The lower panel displays the logarithm of the abundance (nucleon fraction) versus time. The upper panel shows the evolution of the energy generation rate ϵ, the radiation loss L/M, and the neutrino cooling s_ν (here near zero).

The cosmological abundance by number of D and ^{3}He are comparable to that of CNO catalysts in Population II stars, so that the burning of these isotopes to carbon would help solve the problem of the deficiency of catalysts. The first stage of burning destroys D, and converts some ^{7}Be to ^{7}Li; this causes a rapid thermal adjustment (not shown).

Then a steady increase in temperature begins, in which all the light nuclei (below ^{12}C) are destroyed, excepting only H and ^{4}He. Essentially none of these nuclei become potential catalysts; they burn to ^{4}He. This causes the weak "blip" in ϵ at $t \approx 3 \times 10^{11}$ s.

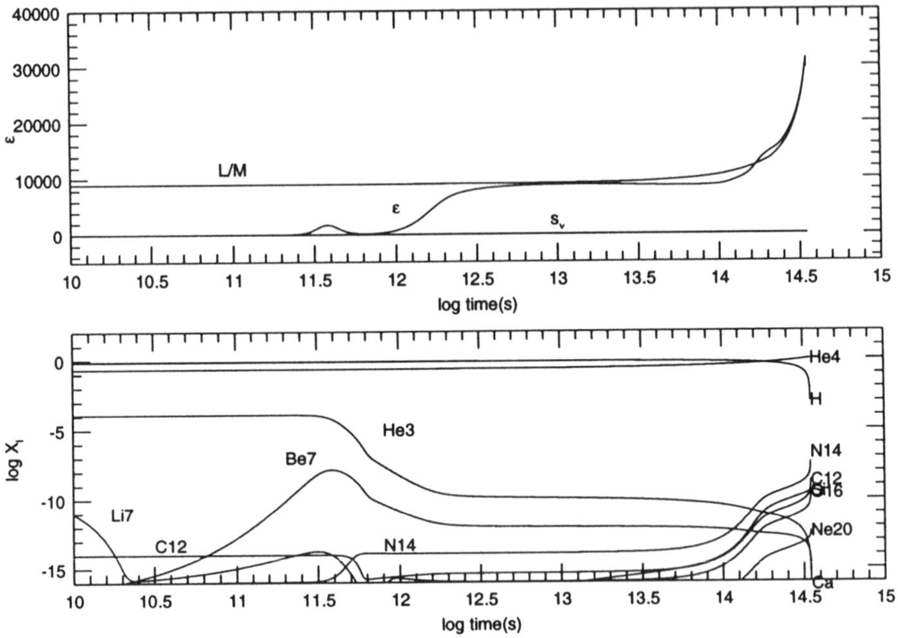

Figure 3. Evolution of Abundances ($S_\gamma = 1$).

4.2. FORMATION OF CNO CATALYSTS

As suggested by the simple analysis above, when the temperature approaches 10^8 K, the triple-alpha reaction begins to produce ^{12}C, which begins a high temperature CNO cycle. Note the rapid rise of ^{14}N, ^{12}C, and ^{16}O, in Figure 3. More surprising is the rise of Si isotopes, which approach a nucleon fraction of 10^{-9}.

The burning proceeds steadily at these temperatures, consuming hydrogen. The abundances after hydrogen burning are shown in Figure 4. Stable nuclei are shown as solid symbols, radioactive by open ones. After ^{1}H and ^{4}He, ^{14}N is most abundant, with a nucleon fraction $\approx 10^{-7}$ (the solar abundance of this nucleus is about 10^{-3}). In this high temperature hydrogen burning, proton reactions produce nuclei up to Ar, although their abundance is low (10^{-9}). Several radioactive nuclei are produced (^{13}N, ^{14}O, ^{15}O, ^{17}F, ^{18}F, (^{22}Na, and ^{26}Al).

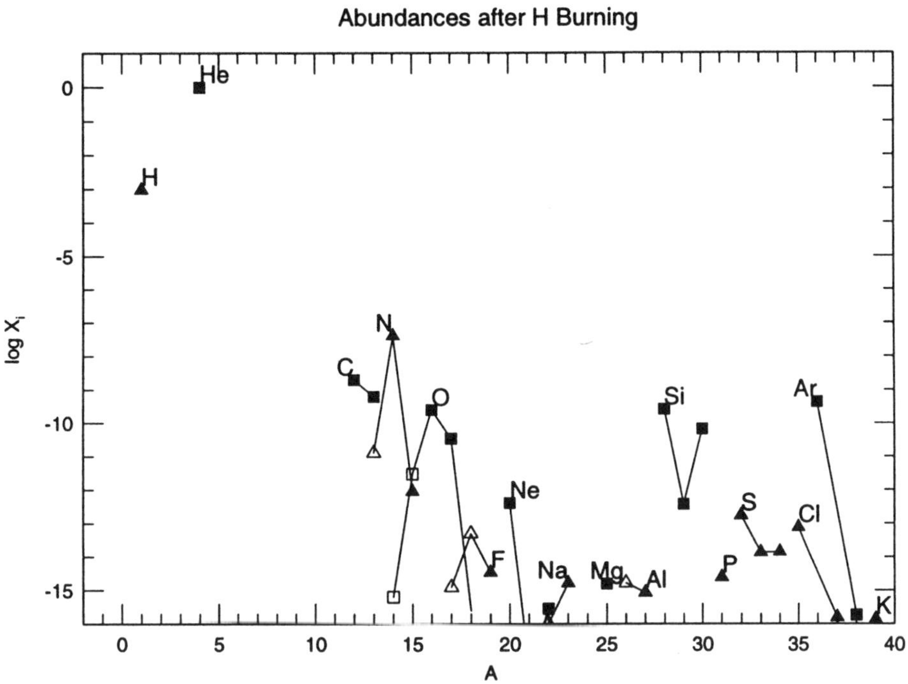

Figure 4. Abundances after Hydrogen Burning.

4.3. CHANGE IN NEUTRON EXCESS

When the H abundance falls below about 10^{-4}, helium burning begins to dominate, producing more ^{12}C and ^{16}O, and some ^{20}Ne as helium begins to be depleted.

With the onset of carbon burning, the production of free protons drives production of ^{14}N, which is then consumed as more massive nuclei are formed. Even at the onset of neon burning, the abundance of iron peak nuclei is tiny: only 10^{-21} or so.

The neutron excess $\eta = (N-Z)/A$ rises steadily during carbon burning. For the solar abundance pattern the neutron excess is $\eta_\odot \approx 1.5 \times 10^{-3}$. For Population II it is smaller by about a factor of 10^{-2}, or $\eta_{pop.II} \approx 1.5 \times 10^{-5}$. At the beginning of carbon burning, $\eta = 1.0 \times 10^{-7}$, and increases to $\eta = 3.5 \times 10^{-4}$ at carbon consumption.

In explosive carbon burning, the abundance of ^{23}Na or ^{27}Al relative to ^{20}Ne or ^{24}Mg is proportional to η (Arnett, 1971). This, along with the contribution to Na and Al production from high temperature hydrogen shell

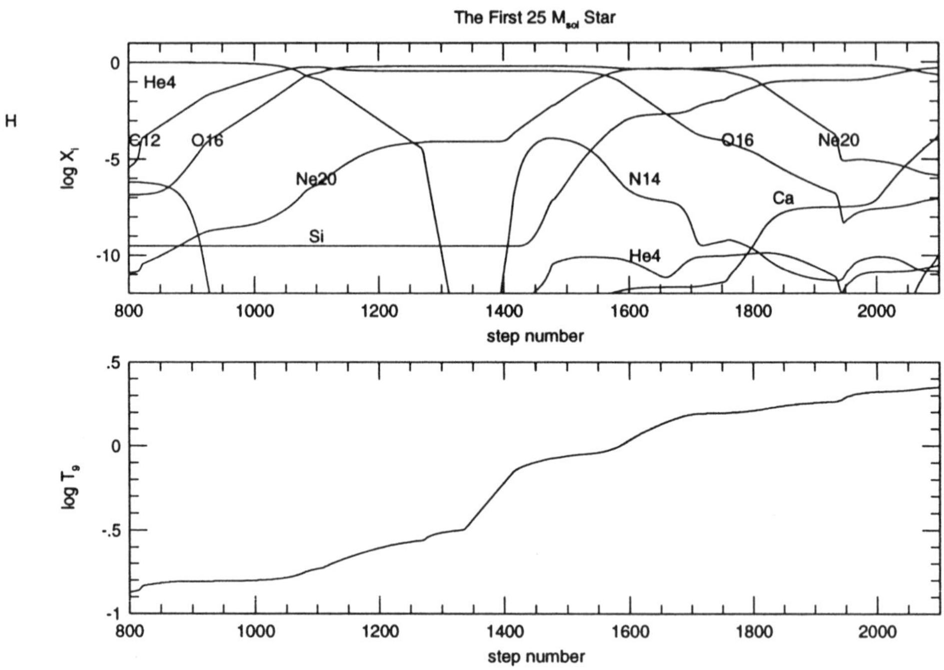

Figure 5. Further Changes in Abundances.

burning, may dominate the most extreme Population II abundances of these elements. If this were the dominant effect, it would imply a deficiency of (Na,Al) to (Ne,Mg) of only a factor of one-fourth.

Hydrostatic neon burning increases η further, to about 5×10^{-4}, or roughly one-third solar. Such layers are oxygen rich, and may undergo explosive oxygen burning. To the extent that most of the neutron excess lies in nuclei with two excess neutrons, such as ^{34}S and ^{38}Ar, the number of neutrons scales as $\sqrt{\eta}$. In quasiequilibrium (explosive oxygen or silicon burning) this causes the ratios of odd/even elements also to scale as $\sqrt{\eta}$. If this were the dominant effect, these ratios would be only about 0.58 solar, which is an interesting notion to test observationally in extreme Population II stars.

5. Production of Ionizing Radiation and Metals

Because essentially the same stars are involved, there is a close connection between the production of ionizing radiation, and the production of carbon and heavier elements ("metals"). This discussion is based upon the formulation in (Arnett, 1996); a critical discussion of the initial mass function (IMF) may be found there.

The total mass burned from hydrogen into helium, for stars from 20 to $100M_\odot$, is

$$\mathrm{M}_\alpha(20, 100) = 0.168\xi_0 X_H, \tag{6}$$

for a Salpeter IMF (the simplest choice). Here ξ_0 is a normalizing constant which can be estimated for the solar neighborhood; it will cancel below. The average energy of photons radiated in the mass range is $< h\nu >= f I_H$, where $I_H = 13.6$ eV. Burning H to He releases $Q = 7$ MeV per nucleon consumed. The total mass of metals produced and returned to space by core collapse supernovae is

$$\mathrm{M}_{COej} = 0.926\xi_0. \tag{7}$$

The ratio of the number of ionizing photons to the number of heavy elements produced in massive stars is then

$$\mathrm{N}_{ion}/\mathrm{N}_z = 1.81(X_H Q A)/(f I_H), \tag{8}$$

where $A \approx 16$ is the mean atomic weight of the heavy elements. This becomes

$$\mathrm{N}_{ion}/\mathrm{N}_z = 1.5 \times 10^7 \ /f. \tag{9}$$

The primary uncertainty may lie in f, which depends upon details of evolution in the HR diagram and mass loss. Several simple attempts at variations in the IMF tended to cancel, leaving this ratio only slowly varying. This extimate may be of value in simulations of the early universe (Gnedin & Ostriker, 1996).

6. Conclusions

This discussion has focused of relatively robust results, that could be simply derived. More detailed stellar evolutionary sequences (one dimensional) were found to be sensitive to the details of convective mixing. Given the results of direct two dimensional simulations of late stellar evolution (Bazan & Arnett, 1998), this is not surprising. A better understanding of stellar mixing will be needed to provide reliable detailed predictions of nucleosynthesis yields. Direct experimental tests (Remington, Weber, Marinak, et al., 1995; Kane, Arnett, & Remington, 1997) of the numerical simulation methods are beginning to give a more secure basis for extended extrapolations of this kind.

7. Acknowledgements

The suggestion of Prof. J. P. Ostriker on the importance of the ratio of production of ionizing radiation to metals is gratefully acknowledged. This work has been supported in part by the National Aeronautics and Space Administration grant NAGW-2798 and National Science Foundation grant AST94-17346.

References

Arnett, D., *Astrophys. J.* **166**, 153, 1971,
Arnett, D., *Supernovae and Nucleosynthesis*, Princeton: Princeton University Press, 1996.
Bazan, G. & Arnett, D., *Astrophys. J.*, 1998, in press.
Clayton, D. D., 1968, *Principles of Stellar Evolution and Nucleosynthesis*, McGraw-Hill, New York.
Gnedin, N., & Ostriker, J. P., *Astrophys. J.***484**, 581, 1996.
Kane, J., Arnett, D., Remington, B., et al., 1997, *Astrophys. J. Letters*, in press.
Kolb, E. W., & Turner, M. S., 1990, *The Early Universe*, Addison-Wesley, Reading, MA.
Remington, B. A., Weber, S. V., Marinak, M. M., et al., *Phys. Plasmas* **2**, 241, 1995.
Schramm, D. N., & Wagoner, R. V., 1979, *Ann. Rev. Nucl. Part. Sci.* **27**, 37.
Wagoner, R. V., Fowler, W. A., & Hoyle, F., 1967, *Astrophys. J.* **148**, 3.

TYPE IA SUPERNOVAE AND CHEMICAL EVOLUTION OF GALAXIES

K. NOMOTO, C. KOBAYASHI, H. UMEDA
Department of Astronomy & Research Center for the Early Universe, School of Science, University of Tokyo, Bunkyo-ku, Tokyo 113-0033, Japan

Abstract. The cosmic/galactic chemical evolutions have been modeled with the early metal enrichment by Type II supernovae (SNe II) and the delayed enrichment of Fe by Type Ia supernovae (SNe Ia). However, the exact nature of SN Ia progenitors have been obscure. Here we present the currently most plausible scenario of the progenitor binary systems of SNe Ia. This scenario involves strong winds from accreting white dwarfs, which introduces important metallicity effects, namely, low-metallicity inhibition of SNe Ia. Resultant predictions for the Galactic/cosmic chemical evolution and the cosmic SNe Ia rate are presented. Another importance of identifying the SN Ia progenitors lies in the use of SNe Ia as a "standard candle" to determine cosmological parameters. To examine whether the "evolution" of SNe Ia with redshift and metallicity is significant, we discuss how the metallicity affects the properties of the C+O white dwarfs such as the C/O ratio, and find the metallicity dependence is rather weak.

1. Introduction

There exist two distinct types of supernova explosions: One is Type II supernovae (SNe II), which are the core collapse-induced explosions of short-lived massive stars ($\gtrsim 8M_\odot$) and produce more O and Mg relative to Fe (i.e., [O/Fe] > 0). The other is Type Ia supernovae (SNe Ia), which are the thermonuclear explosions of accreting white dwarfs (WDs) in close binaries and produce mostly Fe and little O (e.g., Arnett 1996; Ruiz-Lapuente et al. 1997; Nomoto et al. 1994, 1997a). To construct chemical evolution models of galaxies, currently most crucial is to identify the exact progenitor binary systems of SNe Ia.

K. Nomoto and J.W. Truran (eds.), Cosmic Chemical Evolution, 33–46.

Another issue is the application of SNe Ia in cosmology (Branch 1998). Relatively uniform light curves and spectral evolution of SNe Ia have led to the use of SNe Ia as a "standard candle" to determine cosmological parameters, such as the Hubble constant, the density parameter, and the cosmological constant (Perlmutter et al. 1997; Schmidt et al. 1998). However, SNe Ia are not perfect standard candles but show some variations in their absolute magnitudes, light curve shapes, and spectra (e.g., Phillips 1993; Riess et al. 1998). It is important to understand the source of these variations and the dependence on the metallicity in order to use high-redshift SNe Ia for determining the cosmological parameters. The C/O ratio in the SNe Ia progenitors (C-O white dwarfs) may be related to this variation (Höflich et al. 1998; Umeda et al. 1998).

Here we first summarize the effects of stellar mass (M) and metallicity (Z) on the C/O ratio and its distribution in the C-O WDs (§2). We describe currently most plausible model of the SNe Ia progenitors, and introduce a metallicity dependent occurrence of SNe Ia (§3-4). In modeling the galactic/cosmic chemical evolution, this SN Ia model provides a new interpretation of the evolutionary change in [O/Fe] in the solar neighborhood and the SN II-like abundance patterns of the Galactic halo and the damped Lyα systems (§5-6). The metallicity dependent model of SNe Ia provides a testable prediction of cosmic supernova rates (§7).

2. C/O Ratios in White Dwarfs and Metallicity Effects

Umeda et al. (1998) calculated the evolution of intermediate-mass ($3-9M_\odot$) stars for metallicity Z=0.001 – 0.03. Figure 1 shows abundances in mass fraction in the inner core of the 6 $M_\odot$ star for $Y = 0.2775$ and $Z = 0.02$ at the end of the second dredge-up. It is seen that the central part of these stars is oxygen-rich. The C/O ratio is nearly constant in the innermost region, which was a convective core during He burning. Outside this homogeneous region, where the C-O layer grows due to He shell burning, the C/O ratio increases up to C/O $\gtrsim 1$; thus the oxygen-rich core is surrounded by a shell with C/O $\gtrsim 1$. In fact this is a generic feature in all the calculated models. The C/O ratio in the shell is C/O $\simeq 1$ for the star as massive as $\sim 7M_\odot$, and C/O > 1 for less massive stars (Fig. 1).

Figure 2 shows the central C/O ratio in mass fraction, which is not a monotonic function of the metallicity and the ZAMS mass. Two tendencies of the central C/O ratio can be noticed: (1) For a fixed Z, as stellar mass increases, the C/O ratio at the stellar center first increases to the maximum and then decreases. (2) For larger Z, the above maximum C/O ratio is larger. One reason for this complexity is that the C/O ratio is sensitive to the evolutionary change in the convective core size during the later stage

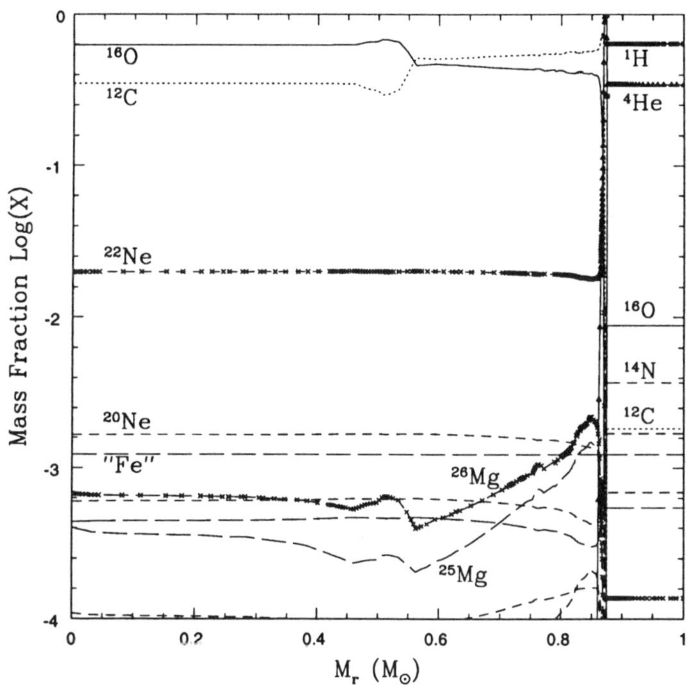

Figure 1. Abundances in mass fraction in the inner core of the 6 $M_\odot$ star for $Y = 0.2775$ and $Z = 0.02$ at the end of the second dredge-up (Umeda et al. 1998).

of He burning. Small differences in the core temperature and opacity lead to a different evolutionary behavior of the convective core, which largely changes the C/O ratio.

If the $3 - 9M_\odot$ star is a primary star in a close binary system, it forms a C+O WD, which in some cases can evolve to a thermonuclear supernova as follows:

(1) If the primary star becomes a red giant (case C evolution), it then undergoes the second dredge-up, forming a thin He layer as seen in Figure 1, and enters the AGB phase. The C+O core mass, $M_{\rm CO}$, at this phase is larger for more massive stars. For a larger $M_{\rm CO}$ the total carbon mass fraction is smaller. When it enters the AGB phase, the star greatly expands and is assumed here to undergo Roche lobe overflow (or a super-wind phase) and to form a C+O WD. Thus the initial mass of the WD, $M_{\rm WD,0}$, at the beginning of mass accretion is approximately equal to $M_{\rm CO}$.

(2) If the primary star becomes a He star (case BB evolution), the second dredge-up in case C corresponds to the expansion of the He envelope even to a red-giant size. The ensuing Roche lobe overflow of the He envelope leads to a white dwarf of mass $M_{\rm WD,0} = M_{\rm CO}$.

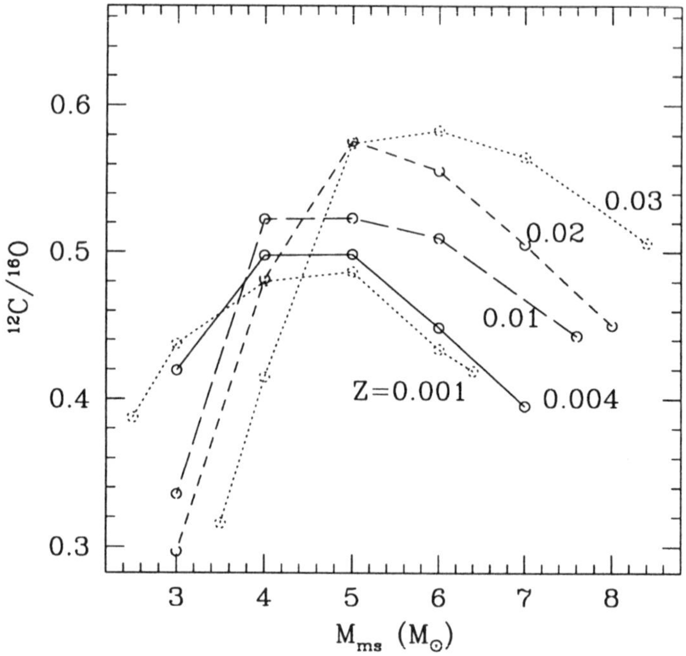

Figure 2. The central C/O ratio in mass fraction as a function of metallicity and stellar mass (Umeda et al. 1998).

(3) After the onset of mass accretion, the WD mass grows through steady H burning and weak He shell flashes. The composition of the growing C+O layer can be approximated as C/O = 1. The WD grows in mass and ignites carbon when its mass reaches $M_{\rm Ia} = 1.378 M_\odot$ (Nomoto et al. 1984).

Figure 3 shows the metallicity dependence of the relation between the ZAMS mass ($M_{\rm ms}$) and $M_{\rm CO} = M_{\rm WD,0}$. We note that stars with higher metallicity form smaller C-O cores if they have the same $M_{\rm ms}$.

The WD mass at the onset of mass transfer from the companion star ($M_{\rm WD,0}$) is critically important for the fate of the accreting WD. For $M_{\rm WD,0}$ as low as $\lesssim 0.7 M_\odot$, the WD mass is difficult to reach $M_{\rm Ia} \simeq 1.37 - 1.38 M_\odot$, unless the companion star is massive (and thus young) enough to supply the required amount of mass. For higher $M_{\rm WD,0}$, the WD can become SNe Ia even with a small mass (and thus long-lived) companion.

Figure 4 shows the total mass of ^{12}C ($M(^{12}{\rm C})$) integrated over the WD mass, $M_{\rm Ia} = 1.378 M_\odot$, just before the SN Ia explosion. Here we assume C/O $\sim$ 1 in the outer layer of $M_{\rm CO} \leq M_r \leq M_{\rm Ia}$ which is formed by He shell burning during accretion. If $M(^{12}{\rm C})$ is plotted against $M_{\rm WD,0} = M_{\rm CO}$ rather than $M_{\rm ms}$, the curves for different metallicity are very similar. The total carbon mass fraction just before SN Ia explosion is seen to vary in

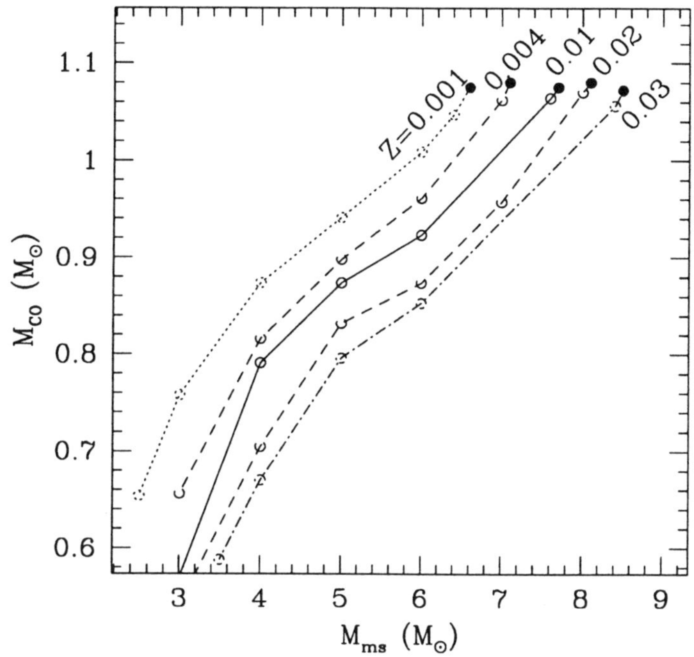

Figure 3. Metallicity dependence of the ZAMS mass (M_{ms}) vs. C-O core mass (M_{CO}) at the end of the second dredge-up. Filled circles indicate that off-center carbon ignition occurs in these model. Hence, the abscissa of these points correspond to the M_{UP} (Umeda et al. 1998).

the range $M(^{12}C)/M_{Ia} = 0.3 - 0.5$.

Such a variation of $M(^{12}C)/M_{Ia}$ may be partially responsible for the observed variations in the brightness and spectra of SNe Ia. Most of previous works on the SNe Ia explosion have used progenitor models with C/O = 1. Only recently Höflich et al. (1998) have started investigation of different C/O for the DD models. The variation of C/O ratio would be important, if it affects the ρ_{tr} in the DD model to induce a large variation in the explosion energy.

The ranges of the variation in the central C/O ratio and $M(^{12}C)/M_{Ia}$ obtained above do not significantly depend on Z for $0.001 \leq Z \leq 0.02$. These ratios exhibit smaller dispersion with respect to metallicity if these ratios are plotted against M_{CO} (Fig. 4). Moreover, WDs with $Z \lesssim 0.002$ need not be considered for SNe Ia progenitors, because such low metallicity WDs blow too weak winds to evolve into SNe Ia as will be discussed in §4. In this sense, the metallicity effects are relatively minor for the same M_{CO}. These would be good for determining cosmological parameters with the use of SNe Ia, because "evolutionary" effects on the SN Ia properties are minor at different red-shifts.

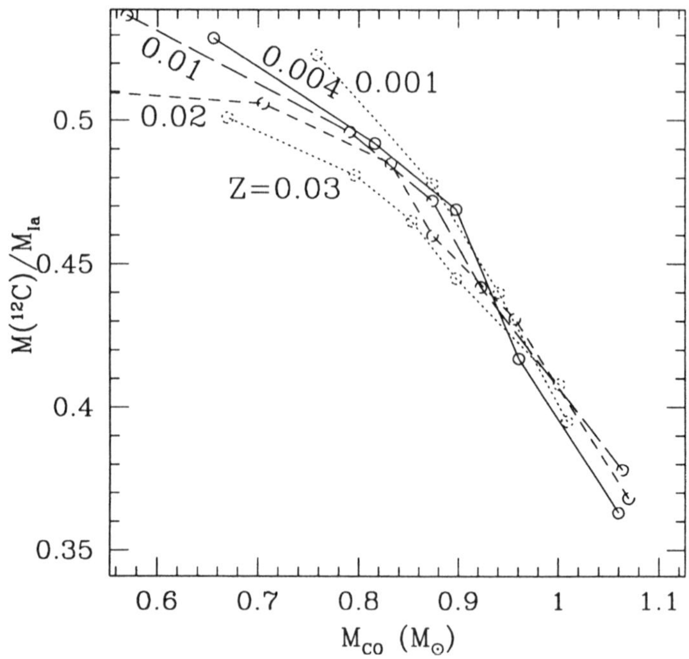

Figure 4. Total ^{12}C mass ($M(^{12}\mathrm{C})$) included in the WD of mass $M_{\mathrm{Ia}} = 1.37 M_\odot$ just before an SN Ia explosion as a function of M_{CO} (Umeda et al. 1998).

3. Progenitors of Type Ia Supernovae

The progenitors of the majority of SNe Ia are most likely the Chandrasekhar (Ch) mass WDs, although the sub-Ch mass models might correspond to some peculiar subluminous SNe Ia (e.g., Nomoto et al. 1997ab). For the evolution of accreting WDs toward the Ch mass, two scenarios have been proposed: One is a double-degenerate (DD) scenario, i.e., merging of double C+O WDs with a combined mass surpassing the Ch mass limit (Iben & Tutukov 1984; Webbink 1984), and the other is a single-degenerate (SD) scenario, i.e., accretion of hydrogen-rich matter via mass transfer from a binary companion (e.g., Nomoto 1982). Though the issue of DD versus SD is still debated, theoretical modeling has indicated that the merging of WDs leads to the collapse to form a neutron star rather than SNe Ia (Nomoto & Kondo 1991).

For the SD scenario, a new evolutionary model has been proposed by Hachisu, Kato, & Nomoto (1996, 1998; hereafter HKN96 and HKN98, respectively). HKN96 have shown that if the accretion rate exceeds a certain limit, the WD blows a strong wind and burns hydrogen steadily to increase the WD mass to M_{Ia}. HKN98 have further invoked the effect of stripping-off of the envelope from the companion by the strong wind, which stabilizes

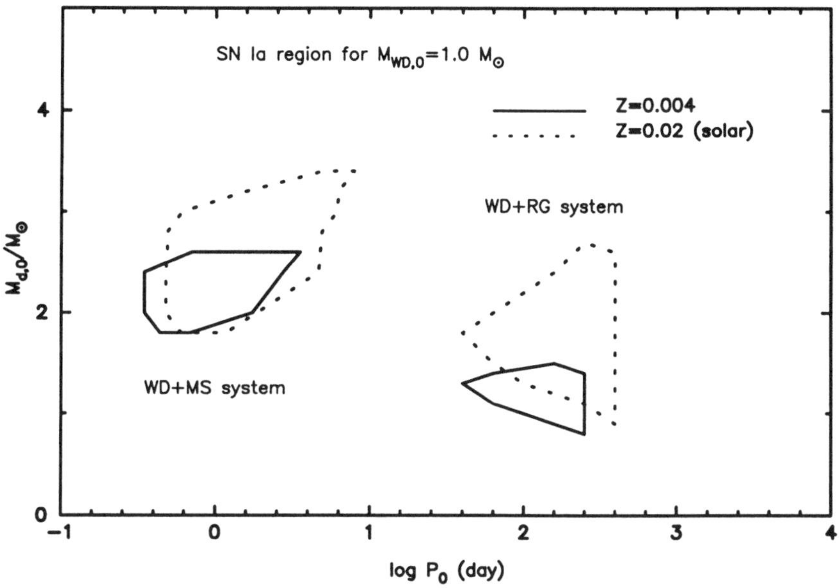

Figure 5. The regions of SNe Ia is plotted in the initial orbital period vs. the initial companion mass diagram for the initial WD mass of $M_{\rm WD,0} = 1.0 M_\odot$. The dashed and solid lines represent the cases of solar abundance ($Z = 0.02$) and much lower metallicity of $Z = 0.004$, respectively. The left and the right regions correspond to the WD+MS and the WD+RG systems, respectively (Kobayashi et al. 1998).

the mass transfer and prevents the system from forming a common envelope. Thus the WD can reach the Ch mass for much wider binary parameter space than found by HKN96, Li & van den Heuvel (1997), and Yungelson & Livio (1998); the allowed parameter space may be large enough to account for the SN Ia frequency.

There are two cases of the companion star: One is a red-giant (RG) with an initial mass of $M_{\rm RG,0} \sim 1 M_\odot$ and an orbital period of tens to hundreds days (HKN96, HKN98), the other is a near main-sequence (MS) with an initial mass of $M_{\rm MS,0} \sim 2 - 3 M_\odot$ and a period of several tenths of a day to several days (Li & van den Heuvel 1997; HKN98).

Figure 5 shows the SN Ia regions in the diagram of the initial orbital period vs. the initial mass of the companion star for the initial WD mass of $M_{\rm WD,0} = 1.0 M_\odot$. In these regions, the accretion from the companion star increases the WD mass successfully through the occurrence of SN Ia. The dashed line shows the case of solar abundance ($Z = 0.02$), while the solid line shows the much lower metallicity case of $Z = 0.004$; the metallicity dependence will be discussed in the next section.

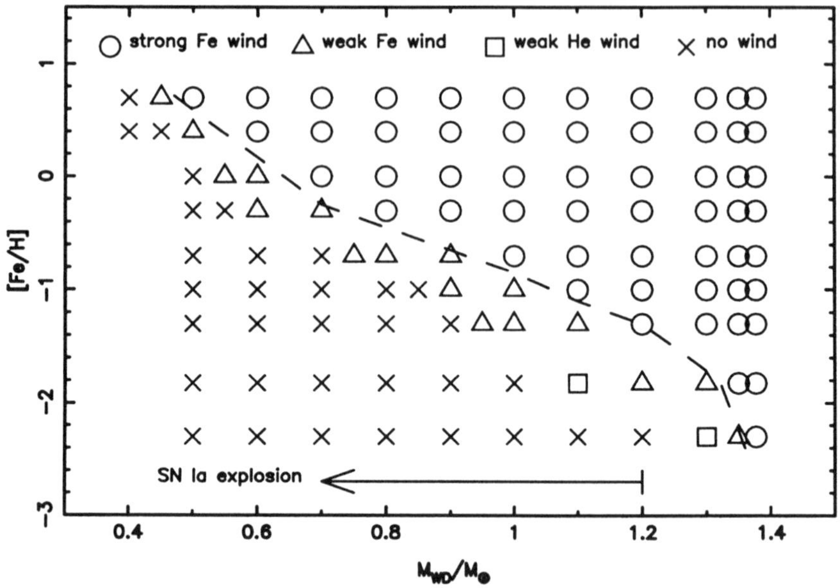

Figure 6. WD mass vs. metallicity diagram showing the metallicity dependence of optically thick winds. The wind is regarded as "strong" if the wind velocity at the photosphere exceeds the escape velocity but "weak" if the wind velocity is lower than the escape velocity. The term of "He" or "Fe" wind denotes that the wind is accelerated by the peak of iron lines near $\log T(\mathrm{K}) \sim 5.2$ or of helium lines near $\log T(\mathrm{K}) \sim 4.6$. The dashed line indicates the demarcation between the "strong" wind and the "weak" wind (Kobayashi et al. 1998).

4. Low-Metallicity Inhibition of Type Ia Supernovae

The white dwarf wind model introduces an important metallicity effect as noted in Kobayashi et al. (1998). The optically thick winds are driven by a strong peak of OPAL opacity at $\log T(\mathrm{K}) \sim 5.2$ (Iglesias & Rogers 1993). Since the peak is due to iron lines, the optically thick winds depend strongly on the metallicity. If the iron abundance of the accreted matter is as low as $[\mathrm{Fe/H}] \lesssim -1$, the WD wind is too weak to increase the WD mass through the Ch mass.

Figure 6 shows the metallicity dependence of the optically thick winds. The strong winds are possible only for the region above the dashed line. The term "weak" implies that the wind velocity at the photosphere does not exceed the escape velocity there, that is, it cannot blow the accreted matter off the WD. For the metallicity as small as $Z = 0.001$, the opacity peak at $\log T \sim 5.2$ is very weak, being smaller than the peak of helium lines at $\log T \sim 4.6$. Then, the wind is driven by the helium line peak rather

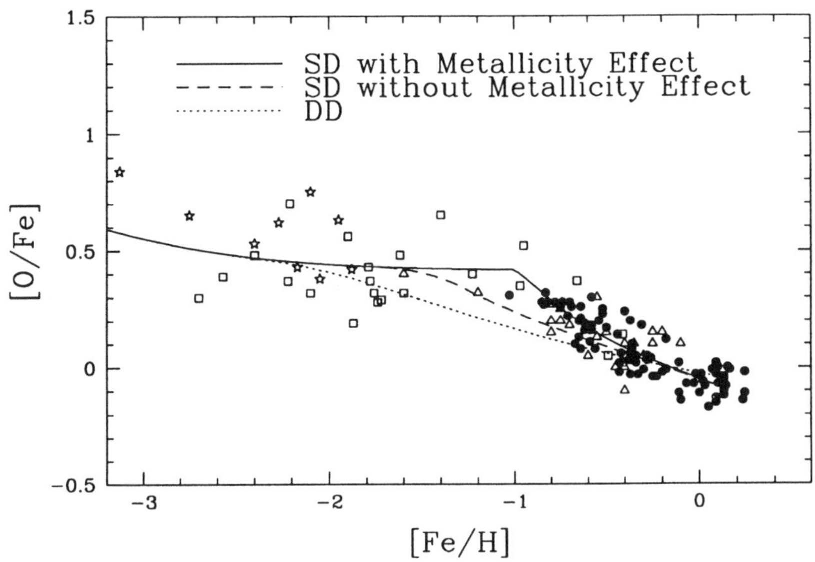

Figure 7. The evolutionary change in [O/Fe] against [Fe/H] for three SN Ia models (Kobayashi et al. 1998). The dotted line is for the DD scenario, and the other lines are for our SD scenario with (solid line) and without (dashed line) the metallicity effect on SNe Ia. Observational data sources: filled circles, (Edvardson et al. 1993); open triangles, (Barbuy et al. 1989); stars, (Nissen et al. 1994); open squares, (Gratton et al. 1991).

than the iron line peak, which we call "He wind" instead of "Fe wind". Since only the initial WD mass of $M_{\rm WD,0} \leq 1.2 M_\odot$ can produce an SN Ia (Nomoto & Kondo 1991), SN Ia events occur only for the progenitors with $[{\rm Fe/H}] \gtrsim -1.1$, which is adopted in our chemical evolution model.

The size of SN Ia regions in Figure 5 clearly demonstrate the metallicity effect, i.e., SN Ia regions are much smaller for smaller metallicity. The initial mass ranges of the companion stars for $Z = 0.004$ are $0.9 M_\odot \lesssim M_{\rm RG,0} \lesssim 1.5 M_\odot$ for the WD+RG system and $1.8 M_\odot \lesssim M_{\rm MS,0} \lesssim 2.6 M_\odot$ for the WD+MS system.

5. The Chemical Evolution in the Solar Neighborhood

The role of SNe II and SNe Ia in the chemical evolution of galaxies can be seen in the [O/Fe]-[Fe/H] relation (Fig. 7: Metal-poor stars with $[{\rm Fe/H}] \lesssim -1$ have $[{\rm O/Fe}] \sim 0.45$ on the average, while disk stars with $[{\rm Fe/H}] \gtrsim -1$ show a decrease in [O/Fe] with increasing metallicity. To explain such an evolutionary change in [O/Fe] against [Fe/H], we use the chemical evolution model that allows the infall of material from outside the disk region. The infall rate, the SFR, and the initial mass function (IMF) are given by Kobayashi et al. (1998), and the nucleosynthesis yields of SNe Ia and II are taken from Tsujimoto et al. (1995).

For the SD scenario, the lifetime of SNe Ia is determined from the main-sequence lifetime of the companion star. The initial mass ranges of the companions are $0.9M_\odot \lesssim M_{\rm RG,0} \lesssim 1.5M_\odot$ for the WD+RG system and $1.8M_\odot \lesssim M_{\rm MS,0} \lesssim 2.6M_\odot$ for the WD+MS system. The low-metallicity inhibition of SNe Ia is introduced for the metallicity of the progenitor systems lower than [Fe/H]$= -1.1$, which is determined from the metallicity dependence of the optically thick wind. For the DD scenario, we adopt the distribution function of the lifetime of SNe Ia by Tutukov & Yungelson (1994), majority of which is $\sim 0.1 - 0.3$ Gyr.

Figure 7 shows the evolutionary change in [O/Fe] for three SN Ia models. The dotted line is for the DD scenario. The other lines are for our SD scenario with (solid line) and without (dashed line) the metallicity effect on SNe Ia. i) In the DD scenario the lifetime of the majority of SNe Ia is shorter than 0.3 Gyr. Then the decrease in [O/Fe] starts at [Fe/H] ~ -2, which is too early compared with the observed decrease in [O/Fe] starting at [Fe/H] ~ -1. ii) For the SD scenario with no metallicity effect, the companion star with $M \sim 2.6M_\odot$ evolves off the main-sequence to give rise to SNe Ia at the age of ~ 0.6 Gyr. The resultant decrease in [O/Fe] starts too early to be compatible with the observations. iii) For the metallicity dependent SD scenario, SNe Ia occur at [Fe/H] $\gtrsim -1$, which naturally reproduce the observed break in [O/Fe] at [Fe/H] ~ -1.

6. Galactic Halo and damped Lyα systems

The low-metallicity inhibition of SNe Ia can provide a new interpretation of the SN II-like abundance patterns:

1) The Galactic halo is a low-metallicity system with [Fe/H] $\lesssim -1$, and it has an abundance pattern of genuine SN II origin, i.e., the overabundances of α-elements relative to Fe as [α/Fe] > 0 (Wheeler, Sneden & Truran 1989). However there exist age differences of several Gyrs among the clusters as well as field stars. Since the shortest lifetime of SNe Ia is ~ 0.6 Gyr for the MS+WD close binary systems, SN Ia contamination would be seen in [α/Fe] if there were no metallicity effect on SNe Ia. This apparent discrepancy between the age difference and the high [α/Fe] can be resolved by the low-metallicity inhibition of SNe Ia.

2) A similar interpretation also holds for DLA systems. The DLA systems observed at $0.7 < z < 4.4$ have [Fe/H]$= -2.5$ to -1 and indicate [α/Fe] > 0 (Lu et al. 1996). The age-metallicity relation in DLA systems suggested by Lu et al. (1996) implies that DLA systems have grown through a common chemical history spanning over several Gyrs. If so, the SN II-like abundance

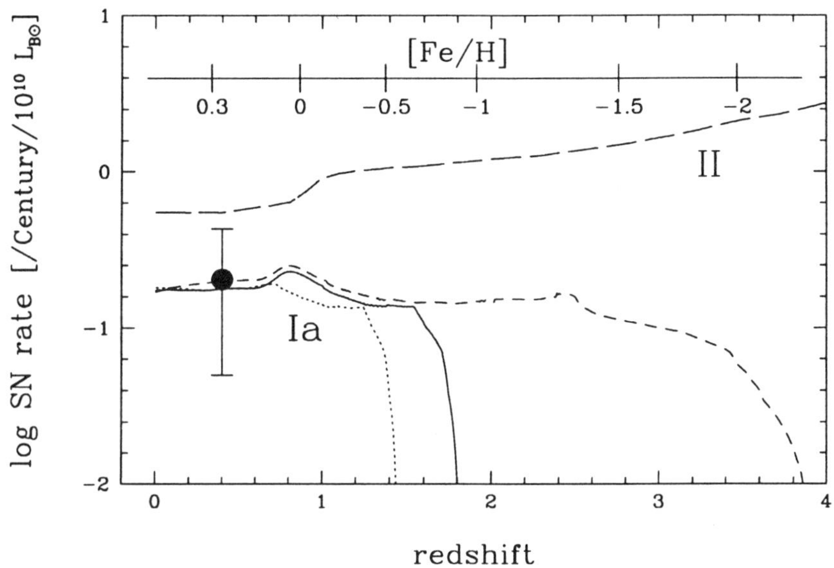

Figure 8. The cosmic supernova rate per $10^{10}L_\odot$ per century (SNu). The long-dashed line is for SNe II and the other lines for SNe Ia with (solid and dotted lines) and without (dashed line) the metallicity effect. The correction of the dust extinction is included in the SFR (Pettini et al. 1998) except for the dotted line which shows the SN Ia rate without this correction (Madau et al. 1996). The filled circle is the observed SN Ia rate at $z \sim 0.4$ (Pain et al. 1996). The iron-abundance scale in the abscissa is calculated with the metallicity effect on SNe Ia.

pattern in DLA systems needs the introduction of the metallicity dependent SN Ia rate to avoid the contamination of SN Ia products.

7. Cosmic Supernovae Rate

SNe Ia have been discovered up to $z \sim 1.2$ by the Supernova Cosmology Project and the High-z Supernova Search Team. They have given the SN Ia rate at $z \sim 0.4$ (Pain et al. 1996) but will provide the SN Ia rate history over $0 < z < 1.3$. With the Next Generation Space Telescope, both SNe Ia and II will be observed through $z \sim 4$. In a theoretical approach, the cosmic SN Ia rate as a function of redshift has been constructed for a cosmic star formation rate (SFR). In this paper, we make a prediction of the cosmic SN Ia rate as a composite of different types of galaxies which have gone through different chemical enrichment.

First, we predict the cosmic supernova rate history corresponding to the observed cosmic SFR (e.g., Madau et al. 1996) with the correction of the dust extinction (Pettini et al. 1998). The photometric evolution is calculated with the spectral synthesis population database taken from Kodama &

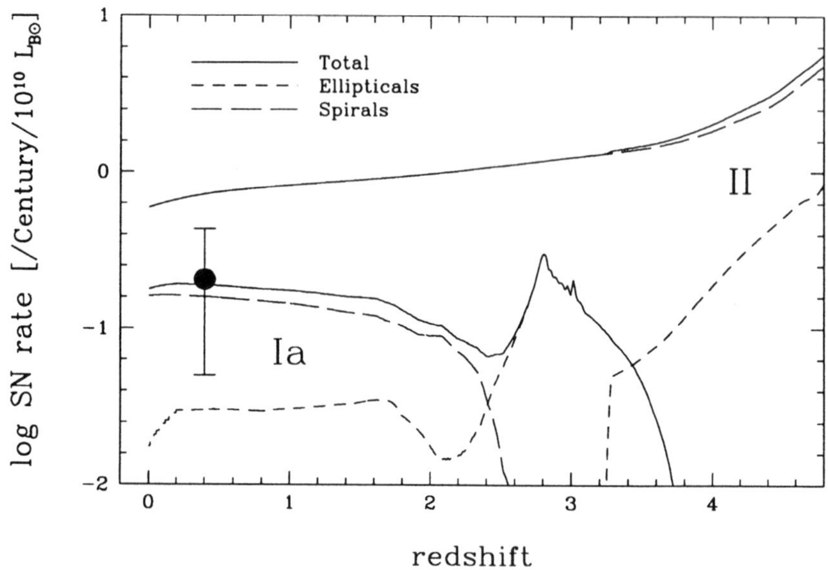

Figure 9. The cosmic supernovae rate in SNu as a composite of ellipticals and spirals. The upper three lines show SN II rates, the lower three lines show SN Ia rates. The solid-line, the short dashed-line, and the long dashed-line are for total, ellipticals, and spirals, respectively.

Arimoto (1997). We adopt $H_0 = 50$ km s^{-1} Mpc^{-1}, $\Omega_0 = 0.2$, $\lambda_0 = 0$, and the redshift at the formation epoch of galaxies $z_f = 5$. We determine the initial comoving density of gas $\Omega_{g\infty}$ to reproduce the present gas fraction $\Omega_{g0} = 5 \times 10^{-4}$.

Figure 8 shows the cosmic supernova rate per $10^{10} L_\odot$ per century (SNu). The long-dashed line is for SNe II and the other lines for SNe Ia with (solid and dotted lines) and without (dashed line) the metallicity effect. The dotted line shows the SN Ia rate without the correction of the dust extinction in the SFR (Madau et al. 1996).

If we do not include the metallicity effect, the SN Ia rate is almost flat from the present to higher redshift, and decreases toward the formation epoch of galaxies. If we include the metallicity effect, the SN Ia rate drops at $z \sim 1.4 - 1.8$, where the iron abundance of the gas in the universe is too low (i.e., [Fe/H] $\lesssim -1$) for the progenitors of SNe Ia to make explosions. The redshift where the SN Ia rate drops is determined by the speed of the chemical enrichment, which depends on the effect of dust extinction on the cosmic SFR, cosmology, galaxy formation epoch, and the initial gas density. Taking into account these uncertainties, the break in the SN Ia rate occurs at $z = 1 - 2$.

In fact, galaxies being responsible for the cosmic SFR have different heavy-element enrichment timescale, so that we calculate the cosmic chemical evolution as a composite of ellipticals and spirals (S0a-Sa, Sab-Sb, Sbc-Sc, and Scd-Sd). We assume that elliptical galaxies are formed by a single star burst and stop the star formation after the supernova-driven galactic wind (Kodama & Arimoto 1997), while spiral galaxies are formed by a relatively continuous star formation. The epoch of the end of star formation in ellipticals is assumed to be 1 Gyr, which correspond to the redshift of $z \sim 3$. The infall rate and the SFR are given by Kobayashi et al. (1998). We then combine the contribution of ellipticals and spirals with the relative mass ratio among the types. The predicted cosmic SFR have the little smaller slope than the observed cosmic SFR from the present to the peak at $z \sim 1.4$ and the excess at $z \gtrsim 2$ corresponding to the SFR of ellipticals which may be hidden by the dust extinction.

Figure 9 shows the cosmic supernova rate (solid-line) as the composite of ellipticals (short dashed-line) and spirals (long dashed-line). The upper three lines show the SN II rates and the lower three lines show the SN Ia rates. The SN Ia rate in spirals decreases at $z \sim 2.5$ because of the low-metallicity inhibition of SNe Ia. We can find many SNe Ia at $z > 3$ in ellipticals, where the chemical enrichment takes place so early that the metallicity effect on SN Ia is not effective.

8. Discussion

We discuss the metallicity effects on the properties of the SN Ia progenitors such as the C/O ratio in the WDs. We introduce a metallicity dependence of the SN Ia rate in the Galactic and cosmic chemical evolution models. In this scenario involving a strong wind from WDs, few SNe Ia occur at [Fe/H] $\lesssim -1$. This model successfully reproduces the observed chemical evolution in the solar neighborhood. Following predictions can be made as a test of these metallicity effects.

i) SNe Ia are not found in the low iron abundance environments such as dwarf galaxies and the outskirts of spirals.

ii) The cosmic SN Ia rate drops at $z \sim 1-2$ because of the low-iron abundance, which can be observed with the Next Generation Space Telescope.

iii) At $z > 3$, SNe Ia can be found in ellipticals where the timescale of metal enrichment is sufficiently short, while the SN Ia rate in spirals drops at $z \sim 2.5$ due to the low-iron abundance.

This work has been supported in part by the grant-in-Aid for COE Scientific Research (07CE2002) of the Japanese Ministry of Education, Science, Sports, and Culture.

References

Arnett, W.D. 1996, Nucleosynthesis and Supernovae (Princeton: Princeton Univ. Press)
Barbuy, B., & Erdelyi-Mendes, M. 1989, A&A 214, 239
Branch, D. 1998, ARA&A 36, 17
Edvardsson, B. et al. 1993, A&A 275, 101
Gratton, R. G. 1991, in IAU Symp. 145, Evolution of Stars: The Photometric Abundance Connection, ed. G. Michaud & A. V. Tutukov (Montreal: Univ. Montreal), 27
Hachisu, I., Kato, M., & Nomoto, K. 1996, ApJ 470, L97 (HKN96)
Hachisu, I., Kato, M., & Nomoto, K. 1998, ApJ, submitted (HKN98)
Höflich, P., Wheeler, J. C., & Thielemann, F. -K., 1998, ApJ 495, 617
Iben, I. Jr., & Tutukov, A. V. 1984, ApJ-S 54, 335
Iglesias, C. A., & Rogers, F. 1993, ApJ 412, 752
Kobayashi, C., Tsujimoto, T., Nomoto, K., Hachisu, I, & Kato, M. 1998, ApJ 503, L155
Kodama, T., & Arimoto, N. 1997, A&A 320, 41
Li, X. -D., & van den Heuvel, E. P. J. 1997, A&A 322, L9
Lu, L., Sargent, W. L. W., Barlow, T. A., Churchill, C. W., & Vogt, S. S. 1996, ApJS 107, 475
Madau, P., Ferguson, H. C., Dickinson, M. E., Giavalisco, M., Steidel, C. C., & Fruchter, A. 1996, MNRAS 283, 1388
Nissen, P. E., Gustafsson, B., Edvardsson, B., & Gilmore, G. 1994, A&A 285, 440
Nomoto, K., 1982, ApJ 253, 798
Nomoto, K., Iwamoto, K., & Kishimoto, N. 1997a, Science 276, 1378
Nomoto, K., Iwamoto, K., et al. 1997b, in Thermonuclear Supernovae, Eds. P.Ruiz-Lapuente et al. (Dordrecht: Kluwer), 349
Nomoto, K., & Kondo, Y. 1991, ApJ 367, L19
Nomoto, K., Thielemann, F. -K., & Yokoi, K., 1984, ApJ, 286, 644
Nomoto, K., Yamaoka, H., Shigeyama, T., Kumagai, S., & Tsujimoto, T. 1994, in Supernovae, Les Houches Session LIV, ed. S. A. Bludman et al. (Amsterdam: North-Holland), 199
Pain, R., et al. 1996, ApJ 473, 356
Perlmutter, S., et al. 1997, ApJ 483, 565
Pettini, M., Kellogg, M.,Steidel, C.,Dickinson, M., Adelberger, K. L, & Giavalisco, M. 1998, ApJ 508, 539
Phillips, M. M. 1993, ApJ 413, L75
Riess, A.G., Press, W.H., & Kirshner, R.P. 1995, ApJ 438, L17
Ruiz-Lapuente, P., Canal, R., & Isern, J., eds. 1997, "Thermonuclear Supernovae" (Dordrecht: Kluwer)
Schmidt, B. et al. 1998, ApJ 507, 46
Tsujimoto, T., Nomoto, K., Yoshii, Y., Hashimoto, M., Yanagida, S., & Thielemann, F.-K. 1995, MNRAS 277, 945
Tutukov, A. V., & Yungelson, L. R. 1994, MNRAS 268, 871
Umeda, H., Nomoto, K., Yamaoka, H., & Wanajo, S. 1998, ApJ, in press
Webbink, R. F. 1984, ApJ 277, 355
Wheeler, J. C., Sneden, C., & Truran, J. W. 1989, ARA&A 27, 279
Yungelson, L., & Livio, M. 1998, ApJ 497, 168

THE CHEMICAL EVOLUTION OF LIGHT ELEMENTS IN OUR GALAXY

and some implications for "Cosmic Chemical Evolution"

N. PRANTZOS

Institut d'Astrophysique de Paris
98bis, Bd. Arago, 75014 Paris

1. Introduction

Progress in the theory of galactic chemical evolution has been very slow and it is only in the solar neighborhood that observations constrain seriously the parameters of the various models. The history revealed on the basis of these data allows only for a small depletion of deuterium (D), less than a factor of 3 from its pregalactic value (Sec. 2.1). The observational data for the rest of the Milky Way disk are much less constraining for the models. They suggest, however, that a much larger astration (and, hence, D depletion) has taken place in the inner Galaxy; the resulting D gradient, measurable by the future *FUSE-LYMAN* mission, should provide invaluable information as to the past history of the disk (Sec. 2.2). Also, assuming that our Galaxy is a typical spiral, one can calculate the properties of disk galaxies as a function of redshift (in the framework of a given cosmological model) and compare to the observed properties of the extragalactic universe: global star formation rate, gas content and metal abundances in gas clouds. It turns out that D can be considerably depleted in galaxy disks, but only at low redshifts (Sec. 2.3).

Observations of the light element Be and B abundances in halo stars show a surprising linear dependence on metallicity. Despite recent theoretical attempts, this linearity still defies a simple theoretical explanation (Sect. 3), but certainly has important implications for our understanding of cosmic ray acceleration and composition.

K. Nomoto and J.W. Truran (eds.), Cosmic Chemical Evolution, 47–56.

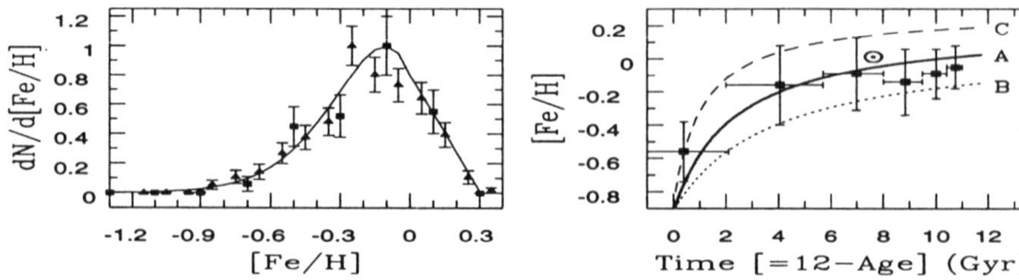

Figure 1. Observables in the solar neighborhood.
a: Metallicity distribution of G-type stars, from 2 recent surveys [25, 34]. The solid line is a fit to the data.
b: Age-metallicity relationship of F-stars, from [5]. The data (182 stars) are binned in groups of 0.2 in log(Age/Gyr); instead of Age, Time=12-Age is plotted here. Metallicities are for stars with galactocentric distances evaluated to 8-9 kpc (i.e. appropriate to local disk) and volume corrected, where the vertical error bars represent 1 σ dispersion in metallicity for each age group (Col. 6 of Tables 14 and 15, respectively, in [5]). The 3 cases A,B and C trace schematically the mean metallicity and the observed dispersion. Notice that simple, one-zone, chemical evolution models always yield a unique age-metallicity relationship and cannot reproduce that scatter without furher assumptions and free parameters.

2. Deuterium evolution in the Milky Way and in high redshifts

2.1. THE PAST HISTORY OF THE SOLAR NEIGHBORHOOD

In the local disk (LD, defined as a cylinder of ~1 kpc radius at a distance $R_{\odot}$=8.5 kpc from the galactic center), the main observables relevant to chemical evolution are ([20, 16]):

• the current surface densities of gas, stars and total amount of matter (~10, 40, and 55 $M_{\odot}$ pc^{-2}, respectively), leading to a current gas fraction of σ_G ~20 %; also, the current star formation rate (SFR), S_0 ~ 1-3 $M_{\odot}$ pc^{-2} Gyr^{-1}.

• the abundances of various elements in the ISM today (Z_0) and 4.5 Gyr ago, the latter presumably represented by the solar system ones ($Z_{\odot}$).

• the oxygen vs. Fe (O-Fe) relationship, showing a steady decline [from O/Fe ~ const. ~3 times solar at [Fe/H]<-1 (i.e. during the halo phase), to O/Fe~solar at [Fe/H]~0], usually attributed to the delayed (~ 1 Gyr) appearance of SNIa, producing 2/3 of the galactic Fe.

• the age-metallicity (Z-t) relationship, traced by the Fe abundance of long-lived, F-type stars.

• the metallicity distribution of long-lived G-type stars (Fig. 1a), showing that very few of them were formed at [Fe/H]<-0.7 (1/5 solar).

The last two observables allow to recover the past SFR history of the

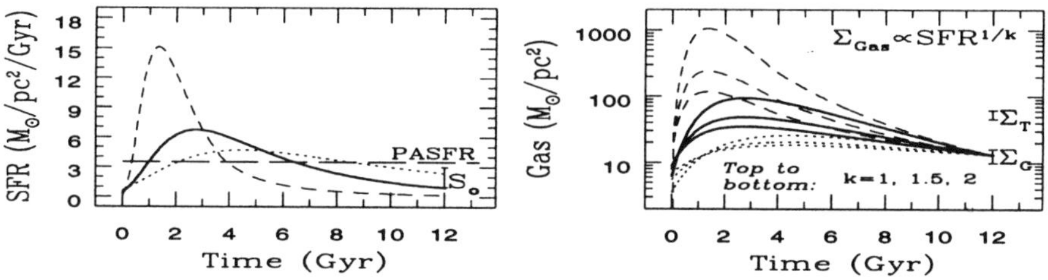

Figure 2. Observationally derived Star Formation Rate in the solar neighborhood. a: The 3 curves correspond to cases A, B and C of Fig. 1b ; they are all normalised to $\int_0^T SFR(t)dt$=42 M$_\odot$ pc^{-2}, with T=12 Gyr. Case C leads to unacceptably low current SFR $\sim$ 0.1 M$_\odot$ pc^{-2} Gyr^{-1}, much lower than the observed S$_0$ $\sim$1-3 M$_\odot$ pc^{-2} Gyr^{-1}. b: Past history of the gas surface density, assuming SFR$\propto$ Σ_G^k. Te 3 types of curves (*solid, dotted, dashed*) correspond again to those of Fig. 1b and 2a. Case C violates the constraint $\Sigma_G < \Sigma_T$ for all plausible values of k.

local disk ([23, 22]), as:

$$SFR \propto \frac{dN_G}{dt} = \frac{dN_G}{d[Fe/H]} \frac{d[Fe/H]}{dt} \tag{1}$$

where the assumption of a constant IMF is necessary for the proportionality between the SFR and the creation rate of G-stars to hold. The scatter in the Z-t relationship does not allow to recover unambiguously the SFR. For illustration purposes, we adopt the 3 curves of Fig. 1b, which have considerably different early and late slopes and cover most of the observed dispersion. When the resulting SFR histories are normalised to the current local star surface density Σ_* over the age T=12 Gyr of the galactic disk ($\int_0^T SFR(t)dt = \Sigma_*$), only cases A and B lead to an acceptable current SFR; case C, as well as all other Z-t relationships with large early slopes, lead to a large early star formation and to too low current SFR (Fig. 2a).

There is another indication against C-type cases. Observations suggest that in disk galaxies SFR$\propto$ Σ_G^k, with k=1-2 (e.g. [9]). Knowing the local SFR history from eq. (1), one can then recover the gas history as Σ_G $\propto$SFR$^{1/k}$, normalising to the current Σ_G. It can be seen from Fig. 2b that C-type cases should be excluded, since they lead to past Σ_G larger than the current total surface density Σ_T.

The above results indicate that the local disk was formed on timescales of many Gyr. Models with slow infall of primordial composition reproduce reasonably well the above constraints. Such models ([21, 31] and Fig. 3) deplete D only by factors $\sim$2-3, at least for reasonable stellar Initial Mass Functions (IMF), like the one of [10]. Notice, however, that the main reason for the small D depletion is the currently large local gas fraction ($\sim$20 %), showing that a rather large fraction of the ISM has not been astrated

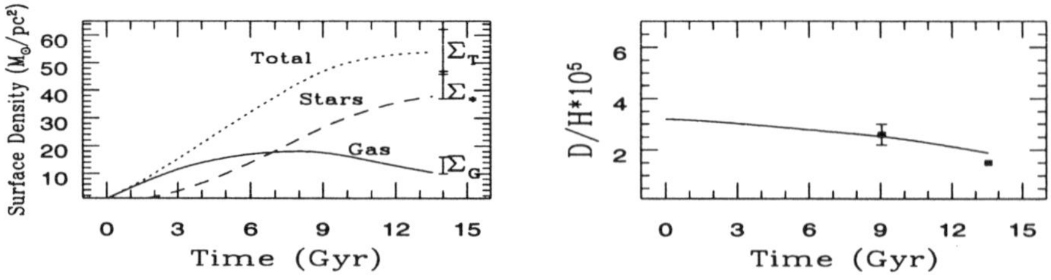

Figure 3. Results of a model for the chemical evolution of the local disk.
a: Evolution of the gaseous, stellar and total mass and comparison of the final results to observations of current local Σ_G, Σ_* and Σ_T, respectively. The adopted SFR is proportional to $\Sigma_G^{1.5}$, while the infall rate is a broad gaussian with $\Delta\tau$=6 Gyr, i.e. the local disk is formed very slowly. Such models reproduce reasonably well the observational constraints of Sec. 2.1 and Figs. 1 and 2a.
b: Corresponding evolution of the D abundance, compared to observations at solar system formation ([8]) and in the local ISM (Linsky, this volume). Only small factors of D depletion (2-3) are obtained in this kind of models, with slow infall of primordial composition.

yet. These results point to low values for pregalactic D/H~3-4 10^{-5}, i.e. compatible with those advocated by Tytler (this volume). Those values also alleviate (but do not solve) the problem of ^{3}He overproduction, which is definitively a stellar nucleosynthesis, not a galactic evolution, problem (see [7] for possible solutions).

2.2. THE EVOLUTION OF THE MILKY WAY DISK

Contrary to the case of the solar neighborhood, the available observations for the Milky Way disk offer information mainly about its current status, not its past history. The main relevant observables are ([20]):

i) The total mass of gas and stars in the disk (~8 10^9 $M_\odot$ and ~5 10^{10} $M_\odot$, respectively); the total current SFR (~3-5 $M_\odot$ yr^{-1}) and the current supernova rates (~2 SNII/century and ~0.4 SNIa/century, respectively, as suggested by observations of external spirals).

ii) The current gas and stellar profiles, showing that the gas fraction profile steeply decreases in the inner disk and suggesting that the SFR efficiency has been larger in those regions than in the outer disk.

iii) The current SFR profile, strongly decreasing outwards, but never proportional to some power of the gaseous profile alone.

iv) The current metallicity profile, traced by oxygen observed in HII regions, planetary nebulae and B-type stars and showing a gradient of d[O/H]~-0.07 dex/kpc ([27, 28]).

These constraints make the Milky Way disk a "testbed" for theories of star formation in disk galaxies. The results of a simple model appear on

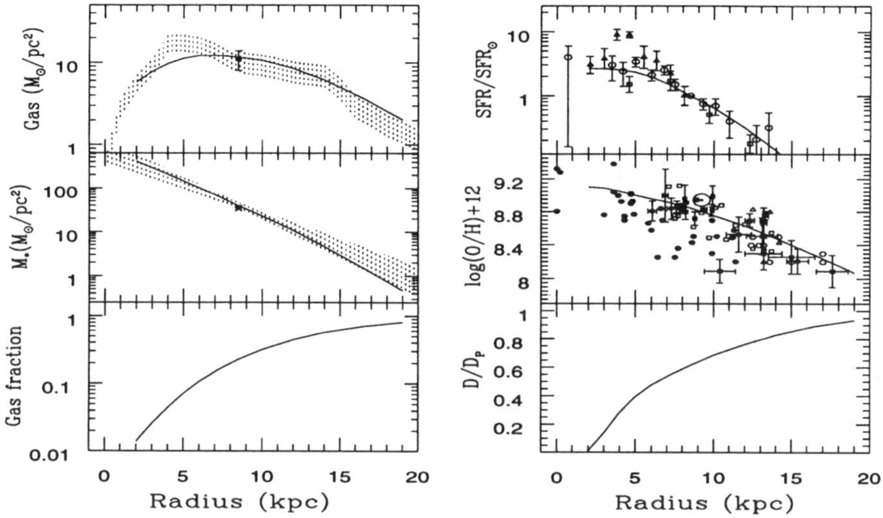

Figure 4. Results of a simple (independent ring) model for the chemical evolution of the Milky Way disk at a galactic age T=13 Gyr, and comparison to observations. The adopted SFR is SFR$\propto \Sigma_G^{1.5}$/R and the adopted infall rate is gaussian in time with $\Delta\tau$=6 Gyr in all the zones.
Left, from top to bottom: final profiles of the surface density of gas, stars and of the gas fraction, respectively. *Solid lines*: model results; *shaded regions* correspond to observations for the disk, and data points at R=8.5 kpc to solar system values.
Right, from top to bottom: current SFR, O and D profiles, respectively. SFR is normalised to its local value and D to its primordial one D_P. Data for O are from HII regions (*open symbols*, from [27, 6]) and B-stars (*filled symbols*, from [28]).

Fig. 4. The adopted SFR $\propto \Sigma_G^{1.5}/R$ is based on the idea that star formation in disk galaxies is induced by the propagation of spiral waves ([33]). The radial variation in the SFR efficiency is the only new parameter introduced in the model, all the others been already fixed by the modelisation of the solar neighborhood (Sect. 2.1). It turns out that with this simple parametrisation the model reproduces reasonably well the constraints (i-iv); it can be used then with some confidence for making further predictions. The most important result of this type of models (in the context of this Conference) concerns the evolution of deuterium. It is found that its final abundance profile shows an important gradient, much more important than (and anticorrelated to) the one of oxygen. The D abundance profile turns indeed to be the most sensitive tracer of the past SFR history of the disk ([21]), especially in the inner regions. Measurements of the D profile with the ISO satellite and the forthcoming FUSE-LYMAN mission should give invaluable information about the past star formation activity in the disk.

2.3. IMPLICATIONS FOR "COSMIC CHEMICAL EVOLUTION"

In recent years, observations of high red-shift systems started revealing some views of the early phases of chemical evolution in the Universe. These observations concern: i) the "global" SFR ([11]); ii) the HI content of the Universe ([29]) and iii) the abundances of various elements in gaseous systems, in the line of sight of quasars ([15, 18]). To these one should include observations of D abundances, although the current status of those observations does not allow to establish a trend as a function of redshift.

Observables (a-c) have prompted several studies of "cosmic chemical evolution" (e.g. [17]). Notice however that, although observables (a) and (b) can be considered as truly "global", observable (c) is rather "local": indeed, it traces the metallicity of only those systems that have still retained some gas at a given redshift. The metals produced e.g. early on in ellipticals and already incorporated in their stars or expelled as ionised gas in the intergalactic medium of galaxy clusters, do not appear in the "census" of (c). The predictions of the "global" model for the evolution of the metallicity cannot then be directly compared to the observations in (c).

Since the only system with data sufficient to constrain its history is the Milky Way, it may be interesting to see how its evolution, modelled in Sect. 2.2 and translated in a cosmological framework, compares to the observables (a-c) of "cosmic evolution". The aim is to see to what extent the Milky Way evolution differs from the "average" one ([22]).

The results are shown in Fig. 5. Milky Way type spirals can account only partially for the observed evolution of the cosmic SFR; this is a reasonable conclusion, since other galaxy types (ellipticals? mergers? earlier type spirals?) are also expected to have a large contribution to the cosmic SFR at $z > 1$. The interesting result of that exercice is that the evolution of Zn in the various disk zones brackets well the observations as a function of the redshift. It may well be indeed, that most of the lines of sight to quasars intercept various parts of (proto)galactic disks. This conclusion is opposite to the one reached in [15], which was based on a comparison of the data to the age-metallicity relationship of the LD only. Notice that for geometrical reasons, the probability of detecting outer disk regions is higher, favouring systematically lower abundances than those spanned during the LD history.

Finally, in the framework of the same model, it turns out that large D depletion may indeed take place in inner galactic disks, but only at low redshifts. D abundances at high redshifts in disk galaxies should always be close to the primordial value. If the observed values differ indeed by the large factors reported in this meeting (Tytler and Lanzetta, respectively), then at least one of the corresponding systems has evolved in a radically different way from a typical spiral.

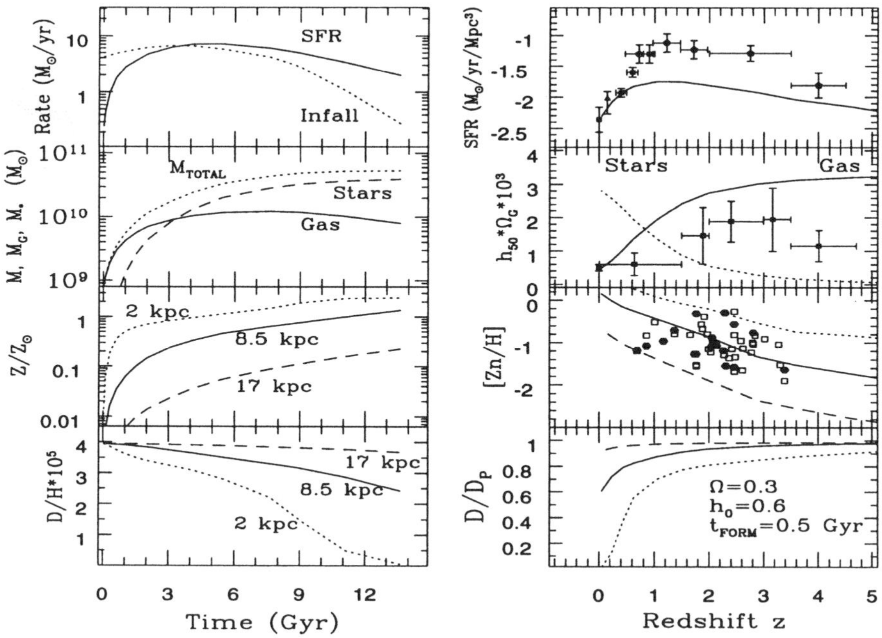

Figure 5. History of the Milky Way disk as a function of time (*left*) and of redshift (*right*), assuming that our galaxy is a typical spiral; a cosmological model with Ω=0.3, h_0=0.6 and galaxy formation starting 0.5 Gyr after the Big Bang is adopted.
Left, from top to bottom: History of a) SFR and infall rate ; b) gaseous, stellar and total mass ; c) overall metallicity in three different zones, at distances of 2, 8.5 and 17 kpc from the galactic center ; d) D abundance in the same regions.
Right, from top to bottom: a) the "cosmic" SFR of disk galaxies,when normalised to the current local value (z=0), does not show the steep observed increase back to z ~1 (although it peaks at z ~1, as observational data do); other galaxy types (ellipticals ?) should account for the discrepancy between theory (*solid line*) and observations (data from [11]) b) Evolution of gas and stars; data for HI gas density from [29]. c) The evolution of Zn/H in various regions of spiral disks (only three are shown here, corresponding to those on the left) brackets well the observed abundances of Zn/H in Lyα absorbers (data from [18, 15]); those systems may well be (proto)galactic disks. d) The corresponding evolution of D shows that considerable depletion may take place in the inner disk regions, but only at low redshifts; abundances measured at high redshifts should be close to the primordial value.

3. Evolution of Be and B in the Galaxy

Thermonuclear reactions in stars cannot synthesize the light isotopes ^{9}Be, ^{10}B, ^{11}B, and ^{6}Li. These, along with part of ^{7}Li, are thought to be produced by spallation reactions of CNO nuclei during the propagation of cosmic rays (CR) in the Galaxy (e.g [24]). Indeed,the observed CR flux, integrated over 10^{10} years of galactic evolution, seems sufficient to produce the solar abundances of B, Be and ^{6}Li (but only ~10% of the solar ^{7}Li, for which some other source, like AGB stars or novae, is required). Adopting CR spectra similar to those observed in near-Earth space (i.e. for energies E>100 MeV/nucleon, since lower E particle fluxes are supressed by the solar wind), the models are able to reproduce the solar system B/Be and ^{6}Li/Be ratios, but produce low ^{11}B/^{10}B~2.5 ratios compared to the observed one (~4). Large fluxes of <100 MeV/nucleon CR particles have been postulated to solve that discrepancy ([13]), the behaviour of the corresponding spallation cross-sections at low energies favouring the production of ^{11}B. More recently, ν-induced nucleosynthesis in SNII has also been suggested as a supplementary source of ^{11}B ([32]).

The evolution of Be and B received considerable attention in the 90ies, since observations of their abundances in halo stars ([26, 3]) showed that: a) contrary to Li, they show no "plateau" value at low metallicity, i.e. no indication for a primordial origin; b) they evolve with a constant ratio of B/Be~10-15, a clear indication of CR origin; c) their relationship w.r.t. Fe is a linear one, over three decades of metallicity. This latter observable is difficult to interpret in the framework of current ideas about CR origin. Indeed, the dominant current component of BeB synthesis (p,α_{CR}+CNO$_{ISM}$, accounting for 70-80% of the total) is of secondary origin, since the abundance of parent CNO$_{ISM}$ nuclei increases with metallicity; despite some theoretical efforts ([26, 19]), this component cannot account for the observed linearity.

Attention was then focused on the currently minor mode of BeB synthesis (CNO$_{CR}$+p,α_{ISM}). If the abundances of CNO$_{CR}$ remain ~const. in time, this mode may produce primary BeB ([3]) and dominate for $Z<Z_\odot/5$. High energy (HE) CR show a source composition similar to the ISM one, i.e. their metallic content is expected, in principle, to evolve in time. However, their observed excess of C and ^{22}Ne is attributed to the contribution of material from the He-burning zones of massive stars, expelled through the Wolf-Rayet (WR) stellar winds ([14]); this material has no time to mix with the ISM before the subsequent SN explosion and may constitute a source of primary BeB. Similar conclusions hold if there exist large scale fluxes of low energy (LE) CR particles in the Galaxy rich in C and O ([2]), a rather strong generalisation of the inferrence made from the recent observations ([1]) of localised 3-7 MeV γ-ray emission from Orion.

Could WR stars provide the source of "metallic" CR for primary Be and B? Unfortunately, the WR phenomenon itself depends heavily on metallicity, as indicated by statistics of WR/O and WC/WR numbers (WC being the sub-type with the He-burning products at the surface) in regions of different metallicities; this is attributed to the reduced effect of radiation pressure on metal-poor stellar envelopes (the main WR formation mecan-

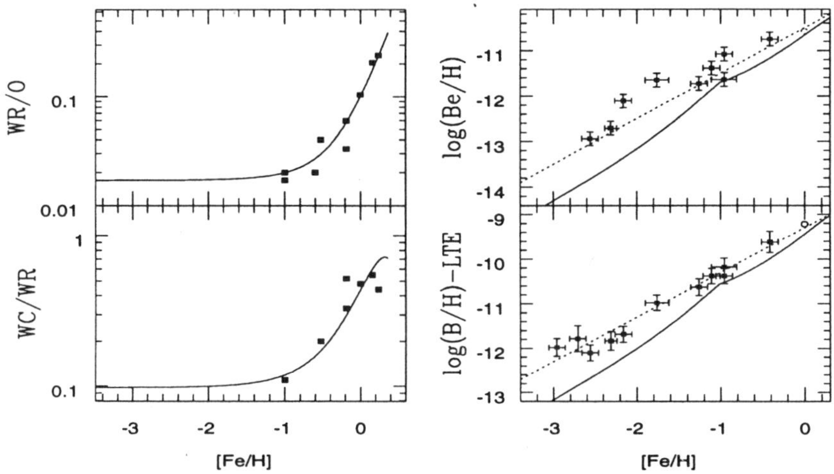

Figure 6. Statistics of WR and WC stars (*left*) and effect of a primary CNO_{CR} component resulting from WC stars on the evolution of Be and B (*right*).
Left: The *solid curve* is a fit to the [Fe/H]>-1 data points (from [12]) and a most optimistic extrapolation at [Fe/H]<-1.
Right: Model results (*solid curve*) compared to *data* (from [4]) and to a slope=1 (*dotted*) curve.

ism for solar metallicity single massive stars). Roche-lobe overflow in binary stars offers an alternative, metallicity independent, mechanism for WR formation but, as seen in Fig. 6, it cannot create more than ~10% of the solar WC/O ratio.

Assuming that WR stars produce primary Be and B and that their frequency evolved in the most "optimistic" way (Fig. 6a), one obtains the evolution of BeB in the framework of a very simple chemical evolution model (Fig. 6b): the linearity is not really reproduced, the deviation between theory and observations becoming more important at low metallicities.

4. Conclusions

The basic points of this short overview may be summarised as follows:

i) Observational data in the solar neighborhood allow in principle to derive the past SFR history of that region. In practice (because of large dispersion in the age-metallicity relationship) they allow only to exclude high early SFRs, favouring a very slow building of the local disk. If this is achieved through infall of primordial composition, only small depletion factors of D (~2-3) are obtained, favouring low primordial values of D.

ii) Observations of the Milky Way disk suggest a high SFR efficiency in its inner regions. Simple toy-models that reproduce reasonably well those observations, predict an important D gradient in the disk; when measured by forthcoming experiments, the D profile will provide invaluable information on the past history of the disk.

iii) The SFR evolution of the Milky Way is presumably smoother than the observed "cosmic" one. However, the range of metal abundances observed in Lyα absorbers corresponds well to the range spanned by the disk of our Galaxy during its evolution; those systems may well be (proto) galactic disks. Finally, according to these simple models, D can be considerably depleted in inner disk zones, but only at low redshifts.

iv) The contribution of WR stars to the CR composition alleviates but does not solve the "enigma" of the linearity of Be and B vs. Fe. This enigma is confirmed by recent observations (Primas, this volume) and its solution will eventually offer new clues as to the origin and acceleration of CR.

References

1. Bloemen H., *et al.*, 1994, A&A Let., 281, L5
2. Casse M., Lehoucq R. & Vangioni-Flam E., 1995, Nature, 373, 318
3. Duncan D., Lemke D. & Lambert D., 1992, ApJ, 401, 584
4. Duncan D., et al., 1997, ApJ, in press
5. Edvardsson B., Andersen J., Gustafsson B., Lambert D., Nissen P., & Tomkin J., 1993, A&A, 275, 101
6. Fich M., & Silkey M., 1991, ApJ, 366, 107
7. Galli D., Stanghellini L., Tosi M. & Palla F., 1997, ApJ, in press
8. Geiss J., 1997, in "Primordial Nuclei and their Galactic Evolution", Eds. N. Prantzos, M. Tosi & R. von Steiger (Kluwer), in press
9. Kennicut R., 1989, ApJ, 344, 685
10. Kroupa P., Tout C., & Gilmore G., 1993, MNRAS, 262, 545
11. Madau P., 1997, in *7th International Origins Conference*, in press
12. Maeder A. & Meynet G., 1994, A&A, 287, 803
13. Menneguzzi M., Audouze J., & Reeves H., 1971, A&A, 15, 337
14. Meyer JP, Elisson D. & Drury J., 1997, ApJ, in press
15. Lu L., Sargent W. & Barlow T., 1996, ApJ Suppl., 107, 475
16. Pagel B., 1997 *Nucleosynthesis and Chemical Evolution*, CUP
17. Pei Y. & Fall M., 1995, ApJ, 454, 69
18. Pettini M., Smith L., King D. & Husntead R., 1997, ApJ, 486, 665
19. Prantzos N., Casse M., & Vangioni-Flam E., 1993, ApJ, 403, 630
20. Prantzos N. & Aubert O., 1995, A&A, 302, 69
21. Prantzos N., 1996, A&A, 310, 106
22. Prantzos N. & Silk J., 1997, in preparation
23. Rana N., 1991, ARAA, 29, 129
24. Reeves H., 1994, Rev. Mod. Phys., 66, 193
25. Rocha-Pinto H. & Maciel W., 1996, MNRAS, 279, 447
26. Ryan S., et al., 1992, ApJ, 388, 184
27. Shaver P., McGee R., Newton L., Danks A. & Pottasch S., 1983, MNRAS, 204, 53
28. Smartt S. & Rolleston W., 1997 ApJ Let., 481, 47
29. Storri-Lombardi L., McMahon R. & Irwin M., 1996, MNRAS, 283, 79
30. Tinsley B., 1980, Fund. Cosm. Phys., 5, 287
31. Tosi M., Steigman G., Matteucci F. & Chiapinni C., 1997, ApJ, submitted
32. Woosley S. et al. 1990, ApJ, 356, 272
33. Wyse R. & Silk J., 1989, ApJ, 379, 700
34. Wyse R. & Gilmore G., 1995, AJ, 110, 2771

NUCLEOSYNTHESIS IN RED GIANT STARS

NAMI MOWLAVI
Geneva Observatory
CH-1290 Sauverny, Switzerland

Abstract. The production of elements from helium-3 to fluorine in low- and intermediate-mass stars is reviewed and compared to chemical abundances observed at the surface of both red giant branch and asymptotic giant branch stars. It is highlighted that, while the trends predicted by standard models are generally well confirmed, many chemical abundances observed at the surface of red giants require the operation of non-standard mixing in the stellar interior. In addition, chemical abundance predictions from presently available asymptotic giant branch models further suffer from the uncertainties affecting the third dredge-up phenomenon, the source of neutrons and the hot bottom burning process.

1. Introduction

The progenitors of the red giant stars considered in this paper are those with masses M less than 6 to $8\,M_\odot$[1] which end their life as white dwarfs (WDs). Considering that their age must be lower than that of the Galaxy, their initial mass is moreover restricted to $M \gtrsim 0.8\,M_\odot$.

The different phases of the evolution of these low- and intermediate-mass (LIM) stars are summarized in [57], to which I refer for a general introduction to the present paper. Let me recall that the surface abundances of LIM stars are altered by several *dredge-up episodes* during which the convective envelope penetrates into the deep regions and mixes to the surface matter processed by nuclear burning. The so-called 'first' and 'second' dredge-ups bring to the surface ashes from hydrogen-burning nucleosynthesis. The first dredge-up (1DUP) occurs after the main sequence phase when the star enters the red giant branch (RGB) phase, and the second dredge-up after the

[1]The case of stars more massive than 6 to $8\,M_\odot$ is reviewed by D. Arnett and K. Nomoto in this volume.

K. Nomoto and J.W. Truran (eds.), Cosmic Chemical Evolution, 57–69.

core He-burning phase (but only for stars more massive than $\sim 4\,M_\odot$) when the star enters the asymptotic giant branch (AGB) phase. In contrast, the so-called 'third' dredge-ups (3DUPs) have the distinctive characteristics of bringing to the surface, in addition to the ashes from the H-burning shell, products from helium burning. These 3DUPs occur after thermal instabilities (called 'pulses') which develop recurrently in the He-burning shell of AGB stars. Finally, the abundances at the surface of AGB stars can further be altered directly by nucleosynthesis operating *in* the envelope when the temperature at its base increases above 50×10^6 K. This scenario is called *hot bottom burning* (HBB).

In the context of cosmic chemical evolution, LIM stars are important actors with respect to elements from He to Al and to the main component of the s-process elements (which are synthesized by neutron capture on iron and heavier elements, see [3]). This derives from three factors: *(a)* the rich nucleosynthesis processes occurring in these stars, and especially during the AGB phase ([57]); *(b)* the high mass loss rates characterizing their late phases of evolution, through which a large fraction of the stellar material is released into the interstellar medium (from about 40% for a $1\,M_\odot$ star to more than 80% for LIM stars with $M \gtrsim 4\,M_\odot$); *(c)* the initial mass function which favors the formation of LIM stars over massive ones in stellar populations.

Ample observational evidences of the nucleosynthesis operating in LIM stars are available in the literature (e.g. [78]). In this paper, I review the status of nuclides from ^{3}He to ^{19}F regarding their nucleosynthesis in LIM stars. The cases of ^{7}Li and ^{23}Na are discussed in [59]. The surface abundance predictions from 'standard' stellar calculations are briefly presented in Sect. 2. Section 3 then focuses on the light element ^{3}He, Sect. 4 on the CNO elements, and Sect. 5 on fluorine. For each of these nuclides, their nucleosynthesis path is recalled and their predicted abundances in red giants and AGB stars compared to observations. Conclusions and final remarks are drawn in Sect. 6.

2. Standard model predictions

'Standard' stellar model calculations refer to those using the Schwarzschild criterion to define the limits of convection boundaries, with *no other* mixing process taken into account, such as overshooting, diffusion or meridional circulation. The surface abundances predicted by such calculations after the 1DUP are widely documented in the literature, at least for the main nuclides (e.g. [23, 76, 86, 17, 31]). They are summarized in Fig. 1 for Pop. I stars from 0.8 to $10\,M_\odot$.

Chemical yields from AGB models are more delicate to predict. Full

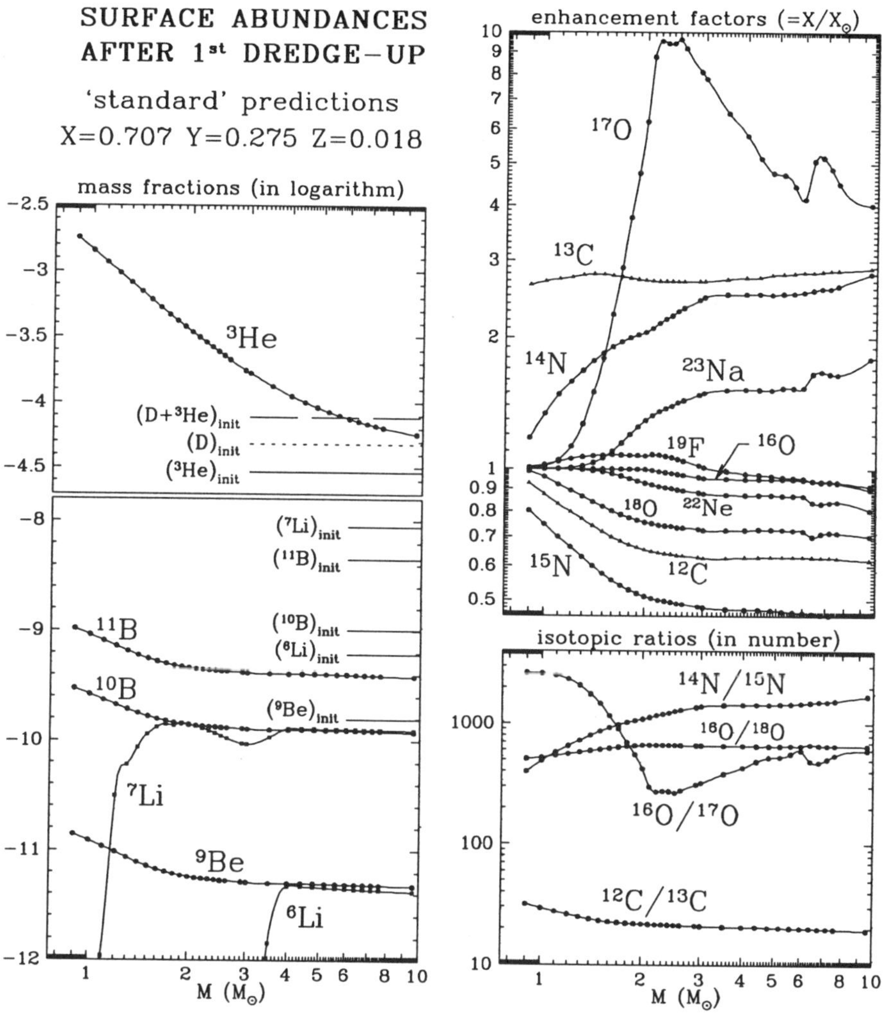

Figure 1. Abundance of elements from ^{3}He to ^{23}Na predicted at the surface of Pop. I red giant stars by standard (i.e. without any extra-mixing beyond the Schwarzschild boundaries for convection) models. The abundances of the Mg and Al isotopes, and of other nuclides not reported on the figures, remain essentially solar during the first dredge-up of Pop. I model models. The stellar evolution code is described in Mowlavi (1995,[56]), modified, among others, to use the OPAL equation of state [73], and the Canuto-Goldman-Mazzitelli prescription for convection [22] (with $K_0 = 1.7$ and $\alpha^* = 0.09$). All models are evolved from the pre-main sequence 'Birth Line' (see e.g. [11]), obtained by accreting $10^{-5}\,M_\odot y^{-1}$ on an initial $0.7\,M_\odot$ model. The p-capture reaction rates are identical to those used in [60], except for ^{17}O(p,α)^{14}N and ^{17}O(p,γ)^{14}N which are from [12], ^{20}Ne(p,γ)^{21}Na, ^{21}Ne(p,γ)^{22}Na, ^{22}Ne(p,γ)^{23}Na, ^{23}Na(p,γ)^{24}Mg and ^{23}Na(p,α)^{20}Ne from [32], and ^{22}Na(p,γ)^{23}Mg, ^{25}Mg(p,γ)^{26}Al and ^{26}Al^g(p,γ)^{27}Si from [26]. The reader is referred to [4] for a discussion on the uncertainties affecting the H-burning reaction rates.

stellar evolutionary calculations do exist for that phase, and indeed provide precious information on the abundances expected at the surface of AGB stars ([14, 88, 16, 13, 85, 35]). However, these standard models have many difficulties to predict self-consistently surface abundances compatible with observational data. For example, AGB models fail to reproduce the observed properties of C stars, such as their luminosity function (cf. the *'C star mystery'*, [49]). The operation of the 3DUP mechanism itself was a puzzling problem to stellar modelers until recently. The presence of s-process elements at the surface of S, SC and C stars presents another difficulty (cf. the *'S star mystery'*, [74]). Fortunately, the situation looks more promising today ([47]). The interested reader is referred to [58] for further discussion on these issues.

Because of these difficulties encountered by current AGB calculations, 'synthetic' AGB models have been used to study the properties of AGB stars. Synthetic models follow the evolution of several stellar characteristics (surface luminosity, chemical composition, ...) based on relations derived from full model calculations (core mass-luminosity relation, ...), but with the use of several parameters (mass loss rate, efficiency of the 3DUPS, mixing parameters, ...). Examples of such synthetic models are found in [68, 55, 87]. These are very useful to determine values of the parameters necessary to reproduce *given* observational properties of AGB stars (such as the luminosity function of carbon stars, [87]), which in turn point out on the shortcomings of present full AGB models. Eventually, however, consistent predictions of AGB yields and properties should be provided by full stellar model calculations.

3. Light elements

The light elements up to boron are easily burned at temperatures of a few 10^6 K (with the exception of H and ^{4}He of course). As a result, they are essentially destroyed by stellar processing. Exceptions, however, are ^{3}He (of importance in relation with Big Bang nucleosynthesis) and ^{7}Li (of interest for Galactic chemical evolution). ^{3}He is discussed in more details in the next paragraphs, while ^{7}Li is developed in [59].

^{3}He: Standard stellar models predict that LIM stars are net producers of ^{3}He (Fig. 1, see also [88, 90]). Two nucleosynthesis paths lead to the production of ^{3}He ([57]): the burning of the initial D during the pre-main sequence phase through $\mathrm{D}(\mathrm{p},\gamma)\,^3\mathrm{He}$, and the p-p chain of H-burning during the main sequence through $\mathrm{H}(\mathrm{p},\mathrm{e}^+\nu)\ \mathrm{D}(\mathrm{p},\gamma)\,^3\mathrm{He}$. First dredge-up then mixes the produced ^{3}He in the envelope, which thereafter remains unaltered according to standard models, except in the most massive AGB stars where it is burned by HBB.

The high ^{3}He abundance observed in the planetary nebula NGC3242 ([70]), of $^3R_{3242} = 50 - 100$ [$^3R \equiv (^3\mathrm{He/H}) \times 10^5$, in numbers], supports the standard model predictions[2]. The same conclusion seems true from measurements in evolved stars ([75, 46, 45]) (but chemical diffusion in those stars may have increased their surface ^{3}He/^{4}He ratio). Moreover, the existence of super lithium-rich stars (see discussion in [59]) implies high ^{3}He abundances in their envelope. Taking into account such ^{3}He production by LIM stars, galactic chemical evolution models predict a continuous increase of the interstellar ^{3}He abundance during the course of Galactic evolution.

This, however, is in obvious contradiction with other ^{3}He abundance determinations: *(a)* $^3R_{HII} = 2.42 \pm 1.45$ in *HII regions*[3] ([69, 7, 9]), considered as representative of the present-day abundance; *(b)* $^3R_{\odot} = 1.48 \pm 0.13$ in the *pre-solar nebula*, i.e. at the time of solar formation, as inferred from several measurements in meteorites (see Appendix A of [36]); *(c)* $^3R_{ISM} = 2.1^{+0.9}_{-0.8}$ in the *local interstellar medium*, as recently measured from isotopic analysis of helium atoms that have entered the Solar System from the surrounding interstellar cloud ([39]). These low values all suggest that LIM stars produce only mildly, if at all, ^{3}He [4]. Moreover, comparison of $^3R_{\odot}$ and $^3R_{ISM}$ reveals that ^{3}He has not been much altered during the past 4.6 Gyr. Actually, abundance determinations of other elements at the surface of red giants strongly suggest that non-standard mixing occurs during the RGB ([77, 82, 24, 48, 25]) and/or the AGB ([16, 89, 1]) phases in at least a category of LIM stars. These processes are also expected to destroy ^{3}He, and galactic chemical evolution models including such destruction are able to account for the above-cited observations (e.g. [66]). But then, the high ^{3}He abundances in planetary nebulae becomes problematic! A solution is to consider that a fraction of LIM stars (10% is suggested by [66]) are still net ^{3}He producers. Abundance determination in a greater sample of planetary nebulae would help to shed some light on this issue (see review by [71]).

[2] Five other planetary nebulae have been analyzed for the presence of ^{3}He. Although only upper limits to their ^{3}He abundance are reported, the authors [8] convincingly argue that ^{3}He is also present in at least one or two of them, with $^3R_{6543} \lesssim 49 \pm 27$ in NGC6543 and $^3R_{7009} \lesssim 69 \pm 27$ in NGC7009.

[3] It is interesting to note that the highest abundance sources are all located in the Perseus arm at ~ 11 kpc from the galactic center, with an average $^3R_{HII,Pers} = 3.32 \pm 1.49$, while the inner galaxy sources have an average $^3R_{HII,in} = 1.45 \pm 0.70$ ([9]).

[4] Starting with a Big Bang Nucleosynthesis value of $^3R_{BBN} \simeq 2$ ([72]) and taking into account the ^{3}He production by LIM stars, galactic chemical evolution models predict an overproduction of ^{3}He with respect to the solar value *even if* only the $\mathrm{D(p,\gamma)}^3\mathrm{He}$ production channel is considered in stellar yields ([67]).

4. CNO elements

^{12}C: This element is destroyed by H-burning in the stellar interior. Its surface abundance thus slightly decreases during the 1DUP (See Fig. 1), but should remain constant, according to standard evolutionary models, during the ascent of the RGB. Yet, observations of ^{12}C among globular cluster giants reveal a progressive decrease of the C/H ratio from the base to the tip of the giant branch by 1.0 dex or more ([82] and references therein). Evidently, some extra-mixing process must occur in those stars *during* their ascent of the RGB.

^{12}C is also the main product of He-burning in AGB stars through the 3-α reactions (~20% in the intershells, in mass fraction). Its mixing in the envelope by the 3DUPs scenario is responsible for the increasing C/O ratio observed in the sequence of M, MS, S, SC and C stars (C/O>1.0 in C stars). As discussed in Sect. 2, AGB models encounter many difficulties in reproducing this scenario consistently and in agreement with observed properties of S and C stars (see [57]). It is also interesting to note that the C/O ratio in C stars does not exceed $\sim$ 1.6, with about 50% of them having 1.0<C/O $\lesssim$ 1.1 (see [54]). It is not clear why this should be so. Apparently, the 3DUP efficiency and/or the total number of pulses experienced during the AGB phase seem to limit the total amount of C dredged-up to the surface during the whole AGB phase. Stellar mass loss may play a key role in this issue. Observations of the C abundance in a larger sample of C stars are welcomed to better constrain the models.

Secondary ^{13}C: Secondary[5] ^{13}C is produced in stellar interiors by p-capture on the initial ^{12}C through $^{12}C(p,\gamma)^{13}C$. First dredge-up then mixes this element in the envelope of red giants. The second dredge-up and 3DUPs do not alter the surface ^{13}C abundance anymore significantly. However, HBB in the most massive AGB stars further transforms ^{12}C into ^{13}C, eventually decreasing the surface ^{12}C/^{13}C ratio down to the CN equilibrium value of 3-4.

The ^{12}C/^{13}C ratio is a good tracer of the evolution of both ^{12}C and ^{13}C. Its predicted value after the 1DUP lies between 20 and 30 (Fig. 1, solar value = 90). This is, in general, in good agreement with the observed ratio in *intermediate-mass* red giants ([31] and references therein). However, many other observed ^{12}C/^{13}C ratios fall below the standard model predictions. These include, for example, those at the surface of low-mass Pop. I field giants (e.g. [31] and references therein), of red giants in Galactic open clusters (e.g. [19, 37, 38]) or in globular clusters (e.g. [83, 20, 18]), and of metal-poor

[5]An element is said to be 'primary' if it can be synthesized in a star of the first generation (i.e. with zero metallicity), and 'secondary' if it needs the presence of some elements heavier than He in the initial stellar chemical composition.

field stars (e.g. [84]). They require some non-standard mechanism (most probably mixing) to operate in low-mass stars (see, e.g. [77, 37, 38]). Two observational works should be mentioned. First, the analysis of the $^{12}C/^{13}C$ ratio in 20 Galactic open clusters as a function of their turn-off mass M_{toff} reveals that this ratio is ~ 25 for $M_{toff} \gtrsim 2.2\,M_\odot$, as expected , but decreases below $2.2\,M_\odot$, reaching ~ 10 at $M_{toff} \simeq 1.2\,M_\odot$[6]([37]). Second, the analysis of the distribution of $^{12}C/^{13}C$ along the giant branch of the open cluster M67 (which has $M_{toff} \simeq 1.2\,M_\odot$) reveals that this ratio is compatible with standard model predictions for stars located at the bottom of the RGB, while those at the tip exhibit anomalously low ratios ([38]). As in the case of ^{12}C in globular clusters, extra-mixing during the ascent of the RGB would explain the $^{12}C/^{13}C$ pattern along the RGB of M67 ([24]). More observations are, however, needed to better characterize the stars which show evidences of such process. This has consequences on abundance predictions of other elements as well, such as ^3He which is expected to be partially (totally?) destroyed by the extra-mixing ([48, 25]).

During the AGB phase, the $^{12}C/^{13}C$ ratio is expected to increase as a result of the surface ^{12}C enhancement brought by the 3DUPs, by a factor up to 4 in C stars having C/O$\simeq$1.2. The ratios observed in AGB stars agree well with these expectations, ranging from $\sim$5 to $\sim$50 in MS, S and SC stars ([79, 30, 29, 80]), and from $\sim$30 to $\sim$70 in C stars ([54, 65, 28]) (see also [52, 51, 5, 6]).

Let me finally mention the special class of C stars called J stars (e.g. [54, 33]), which exhibit very low $^{12}C/^{13}C$ ratios, down to 3-4 ([43]), and which could be identified with AGB stars undergoing HBB ([10]). The high $^{16}O/^{18}O$ ratios observed in those stars support this idea (though some normal C stars have also similar high oxygen isotopic ratios, see sections on ^{16}O, ^{17}O and ^{18}O). However, their N abundances and C/O ratios are not in agreement with HBB expectations. We refer to [43] for further discussion on these objects.

Primary ^{13}C: Primary ^{13}C can be produced in AGB stars in two different ways. The first way is evidenced by the observations of both fluorine and s-process elements in S to C stars ([50]), which are synthesized in the He-burning shell in the presence of neutrons. These neutrons are believed to be produced by $^{13}C\,(\alpha\,,n)\,^{16}O$. A primary source of ^{13}C is necessary because the amount of ^{13}C left over by the H-burning shell in Pop. I stars is too low by a factor of at least 30 in order to match observations of F and s-process elements ([61]). The production of such primary ^{13}C (and ^{14}N) has long been a challenge for AGB modelers ([74]), but looks more promising today

[6]The very low $^{12}C/^{13}C$ ratios of $3-5$ in globular clusters, which have $M_{toff} < 1.0$, enter well into this picture (see Fig. 6 of [24]).

([47]). This primary ^{13}C cannot, however, account for any stellar surface, and thus interstellar, ^{13}C enrichment since it is burned in the He-burning layers before the operation of 3DUP.

Another source of primary ^{13}C (and ^{14}N) is provided in the *envelope* of those AGB stars undergoing HBB ([68]), by H-burning of the primary ^{12}C dredged-up during 3DUP scenarios. Unfortunately, the shortcomings affecting AGB models with respect to 3DUP *and* HBB do not allow presently to provide consistent primary ^{13}C (and ^{14}N) yields from AGB stars.

Secondary ^{14}N: Being the main product of the CNO cycles of H-burning, ^{14}N is largely enhanced after the 1DUP. This is well confirmed by observations, which moreover confirm the expected anti-correlation between N and both C and O (e.g. [81]). The 3DUP, on the other hand, does not affect ^{14}N which is not produced during helium burning. HBB, however, can further enhance the ^{14}N surface abundance in AGB stars.

Primary ^{14}N: Large amounts of primary ^{14}N can be produced by HBB in the envelope of AGB stars *after* primary ^{12}C has been dredged-up from the He-burning layers ([68]). Like for primary ^{13}C, however, no consistent yield of primary ^{14}N is provided yet by current AGB models.

^{15}N: The NO cycles essentially destroy ^{15}N. Its abundance at the surface of red giants is determined by the degree of dilution it undergoes when the envelope deepens in the stellar interior. As a result, the ^{14}N/^{15}N ratio increases during the 1DUP from 270 to 400-1050 in Pop. I stars, depending of the stellar mass (Fig. 1; solar value = 270).

Primary ^{15}N: This nuclide is produced in the He-burning shell of AGB stars (see [60]). Its synthesis results from $^{14}\mathrm{N}\,(\alpha\,,\gamma)\,^{18}\mathrm{F}\,(\beta^+)\,^{18}\mathrm{O}\,(\mathrm{p}\,,\alpha)\,^{15}\mathrm{N}$, where the required protons are provided by $^{14}\mathrm{N}\,(\mathrm{n}\,,\mathrm{p})\,^{14}\mathrm{C}$ and the neutrons by $^{13}\mathrm{C}\,(\alpha\,,\mathrm{n})\,^{16}\mathrm{O}$. As both ^{13}C and ^{14}N required for that synthesis are primary (see section on primary ^{13}C), the resulting ^{15}N is also primary. Nitrogen eventually burns into fluorine through $^{15}\mathrm{N}\,(\alpha\,,\mathrm{p})\,^{19}\mathrm{F}$. According to evolutionary AGB models ([60]), however, the transformation of ^{15}N into ^{19}F is only partial during the first pulses, and becomes more complete as the temperature in the He-burning shell increases. Primary ^{15}N could then be brought to the surface *if* the 3DUP operates early enough along the AGB phase. Again, due to the difficulties of current AGB models to predict both primary ^{13}C *and* the 3DUP scenario (see Sect. 2), no consistent yield for primary ^{15}N (if it is indeed dredged-up to the surface) is available yet.

^{16}O: ^{16}O is partly destroyed by the NO cycle, but its abundance at the surface of red giants is not much altered by the 1DUP (see Fig. 1).

The effect of the 3DUPs on that element is less clear. Determinations of the $^{16}O/^{17}O$ and $^{16}O/^{18}O$ ratios at the surface of MS, S and C stars ([42, 43]) reveal values (600-4000) much higher than those in normal red giants (100-700). The simplest explanation for these high ratios invokes an enhancement of ^{16}O by 3DUP. This scenario seems to be supported by the relative constancy of the $^{17}O/^{18}O$ ratio in those stars (see Table 2 of [27]). From the theoretical point of vue, such a scenario is possible *if* enough ^{16}O is created in the He-burning shell through $^{12}C\,(\alpha\,,\gamma)\,^{16}O$. However, the predicted oxygen composition in the He-burning shell of AGB models is $X(^{16}O)\simeq 1\%$, i.e. comparable to that in the envelope. As a result no surface ^{16}O enhancement is predicted to occur during the 3DUP. And in fact, the ^{16}O abundances measured in C stars are essentially similar to those in M and MS giants ([54]). The oxygen isotopic ratios remain thus an intriguing puzzle for stellar modelers.

^{17}O: ^{17}O is produced in the NO cycle of H-burning, in amounts very dependent on the temperature at which hydrogen burns. Figure 2 of [4] indicates that the maximum ^{17}O production occurs at temperatures around 17×10^6 K. Because temperatures in stellar interiors increase with the stellar mass, a large scatter is expected to characterize the ^{17}O abundances at the surface of red giants as well. Figure 1 indicates that the maximum enhancement of ^{17}O after the 1DUP is found for stars with masses between 2.2 and 2.5 $M_{\odot}$. Concomitantly, the predicted $^{16}O/^{17}O$ ratio after the 1DUP varies from 100 to 600 depending on the stellar mass. The second dredge-up does not alter the ^{17}O abundance much further.

A large scatter in the $^{16}O/^{17}O$ ratio is indeed observed in red giants ([40, 44, 21]), which can be ascribed to differences in stellar masses in agreement with model predictions. It is interesting to note that, in contrast to $^{16}O/^{17}O$, the $^{12}C/^{13}C$ ratios in those red giants fall below the expected values. The extra-mixing required to explain the low carbon isotopic ratios in those stars should thus operate only down to layers between the ^{13}C-rich and ^{17}O-rich zones.

As for the ^{17}O abundance in AGB stars, standard models predict no alteration during a 3DUP, while HBB produces ^{17}O ([16]). The high $^{16}O/^{17}O$ ratios observed in MS, S and C stars, and which could not be explained by a ^{16}O enhancement (see Sect. on ^{16}O), remain thus without any explanation[7]! I refer to [27] for further discussion. More observations of the isotopic ratios in AGB stars are welcomed.

[7]The $^{16}O/^{17}O$ ratios observed in a few SC stars (between 75 and 1200, [30, 29]) are lower than those observed in both S and C stars. If these values are confirmed, they raise a conceptual difficulty since the evolutionary state of SC stars is believed to lie between those of S and C stars.

^{18}O: ^{18}O is destroyed during the NO cycles. As a result, a slight increase in the surface $^{16}O/^{18}O$ ratio is expected to occur after the 1DUP (Fig. 1). This increase, however, is smaller than the observational error bars on that ratio (see, e.g., Fig. 7a of [31]), and no conclusion can be drawn from the comparison of observations with predictions.

In AGB stars, ^{18}O is not expected to be altered by 3DUP according to standard models. The $^{16}O/^{18}O$ ratio should thus remain unchanged[8]. The high $^{16}O/^{18}O$ ratios observed in MS, S and C stars raises difficulties similar to those from $^{16}O/^{17}O$ (see Sect. on $^{16}O/^{17}O$).

^{18}O can also be destroyed by HBB, resulting in an increase of the $^{16}O/^{18}O$ ratio by several orders of magnitude ([16]). Such high values should characterize the ^{13}C-rich J stars (see Sect. on ^{13}C).

5. Fluorine

The observation of large ^{19}F over-abundances in S and C stars relative to those in normal giants ([50]) strongly suggests the production of that element in AGB stars. The existence of a correlation between ^{19}F and s-process elements nucleosynthesis is moreover established on both observational ([50]) and theoretical ([61]) grounds. Stellar evolutionary calculations indeed confirm the production of ^{19}F in AGB models ([34, 60]), but require an additional source of primary ^{13}C. Like the s-process elements thus, fluorine predictions suffer from the inability of current AGB models to produce consistently primary ^{13}C *and* the 3DUP scenario. We refer to the above-cited references for further discussion on the fluorine issue in red giants (see also [62] in this volume).

Let us also remark that secondary ^{19}F could be produced during H-burning by $^{18}O\,(p\,,\gamma)\,^{19}F$ ([4]). Stellar models however, predict only very little enhancement of that element after the 1DUP in LIM stars (Fig. 1; see also discussion in [60]).

6. Conclusions and final remarks

This short review has focused on the discrepancies between the predicted and observed abundances at the surface of both RGB and AGB stars. It is

[8]Actually, ^{18}O is produced by $^{14}N\,(\alpha\,,\gamma)\,^{18}O$ in the He-burning shell, ^{14}N being provided by the ashes of the H-burning shell, and destroyed by $^{18}O\,(\alpha\,,\gamma)\,^{22}Ne$. Evolutionary models show that the destruction is only partial during the first pulses where the temperatures are lower than about 2.2×10^{8} K. As a result, ^{18}O could be mixed to the surface by a 3DUP *if* the latter occurs during the very first pulses. No current AGB model confirms this scenario, though. However, *one* star, the barium star HD101013, presents a $^{16}O/^{18}O$ ratio as low as 60^{+100}_{-30} ([41]). *If* confirmed, it could mean that a (very small) fraction of AGB stars do indeed dredge-up ^{18}O from the He-burning layers.

shown in particular that, while the predicted trends from 'classical' stellar models are well confirmed by the observations, quantitative comparisons put some limitations on the predictive capabilities of such models. Most of the discrepancies suggest that mixing occurs in the stellar interior deeper than that predicted by classical models. In addition, chemical abundance predictions from presently available AGB models further suffer from the uncertainties affecting the third dredge-up phenomenon, the source of neutrons and the hot bottom burning process.

Finally I must mention a new class of objects which have been analyzed during the last decade, and which provide valuable information on the chemical abundances ejected by stars. These are inclusions in primitive meteorites which show anomalous isotopic compositions (i.e. different, often by large amounts, from the relatively homogeneous abundances reported in material of Solar System origin). The currently accepted explanation for these anomalies considers that these grains are of stellar origin, and have been injected in the Solar System during its formation. The fact that SiC grains have also been detected in circumstellar envelopes of carbon stars through their $11.4\mu m$ emission feature (e.g. [53] and references therein) strongly supports that scenario. I refer to [2, 91, 63] for some general reviews on the interstellar grains from primitive meteorites and on their isotopic composition. In particular, it is interesting to note that the analysis of the oxygen isotopic ratios in those grains supposed to have originated from AGB stars leads – again – to the conclusion that extra-mixing must have operated in those stars ([15, 64]). Such studies nicely complement the direct isotopic abundance determinations from the light spectrum of red giants.

Acknowledgment: I thank Drs. Alain Jorissen and Georges Meynet for their careful reading and commenting on the manuscript.

References

1. Abia C., Isern J.: 1997, MNRAS 289, L11
2. Anders E., Zinner E.: 1993, Meteoritics 28, 490
3. Arnould M.: 1991, in 'Evolution of Stars: The Photospheric Abundance Connection' (IAU Symp. 145), eds. Michaud G, and Tutukov A., Kluwer, p 287
4. [9] Arnould M., Mowlavi N., Champagne A.: 1996, in *32nd Liège Int. Astroph. Coll.*, eds. Noels A., Fraipont-Caro D., Gabriel M., Grevesse N., Demarque P., pp 17-29
5. Bakker E.J., Van Dishoeck E.F., Waters L.B.F.M., Schoenmaker T.: 1997, A&A 323, 469
6. Bachiller R., Forveille T., Huggins P.J., Cox P.: 1997, A&A 324, 1123
7. Bania T.M., Rood R.T., Wilson T.L.: 1987, ApJ 323, 30
8. Balser D.S., Bania T.M., Rood R.T., Wilson T.L.: 1997, ApJ 483, 320

[9] postscript files available by anonymous ftp on 'obsftp.unige.ch' in 'pub/mowlavi', or in 'http://obswww.unige.ch/~mowlavi/publications/publications.html

9. Balser D.S., Bania T.M., Brockway C.J., Rood R.T., Wilson T.L.: 1994, ApJ 430, 667
10. Becker S.A., Iben I.Jr.: 1980, ApJ 237, 111
11. Bernasconi P.A.: 1996, A&AS 120, 57
12. Blackmon J.C., Champagne A.E., Hofstee M.A., Smith M.S., Downing R.G., Lamaze G.P.: 1995, Phys. Rev. Lett. 74, 2642
13. Blöcker T.: 1995, A&A 297, 727
14. Boothroyd A.I., Sackmann I.-J.: 1988, ApJ 328, 653
15. Boothroyd A.I., Sackmann I.-J., Wasserburg G.J.: 1994, ApJ 430, L77
16. Boothroyd A.I., Sackmann I.-J., Wasserburg G.J.: 1995, ApJ 442, L21
17. Bressan A., Fagotto F., Bertelli G., Chiosi C.: 1993, A&AS 100, 647
18. Briley M.M., Smith V.V., Lambert D.L.: 1994, ApJ 424, L119
19. Brown J.A.: 1987, ApJ 317, 701
20. Brown J.A., Wallerstein G.: 1989, AJ 98, 1643
21. Campbell B., Lambert D.L., Maillard J.-P.: 1990, PASP 102, 79
22. Canuto V.M., Goldman I., Mazzitelli I.: 1996, ApJ 473, 550
23. Castellani V., Chieffi A., Straniero O.: 1990, ApJS 74, 463
24. Charbonnel C.: 1994, A&A 282, 811
25. Charbonnel C.: 1995, ApJ 453, L41
26. Coc A., Mochkovitch R., Oberto Y., Thibaud J.P., Vangioni-Flam E.: 1995, A&A 299, 479
27. Dearborn D.S.P: 1992, Phys. Reports 210, 367
28. de Laverny P., Gustafsson B.: 1998, A&A, in press
29. Dominy J.F., Wallerstein G.: 1987, ApJ 317, 810
30. Dominy J.F., Wallerstein G., Suntzeff N.B.: 1986, ApJ 300, 325
31. El Eid M.F.: 1994, A&A 285, 915
32. El Eid M.F., Champagne A.E.: 1995, ApJ 451, 298
33. Lloyd E.T..: 1995, Astr.Sp.Sci. 230, 169
34. Forestini M., Goriely S., Jorissen A., Arnould M.: 1992, A&A 261, 157
35. Forestini M., Charbonnel C.: 1997, A&AS 123, 241
36. Galli D., Palla F., Ferrini F., Penco U.: 1995, ApJ 443, 536
37. Gilroy K.K.: 1989, ApJ 347, 835
38. Gilroy K.K., Brown J.A.: 1991, ApJ 371, 578
39. Gloeckler G., Geiss J.: 1996, Nature 381, 210
40. Harris M.J., Lambert D.L.: 1984, ApJ 285, 674
41. Harris M.J., Lambert D.L., Smith V.V.: 1985, ApJ 292, 620
42. Harris M.J., Lambert D.L., Smith V.V.: 1985, ApJ 299, 375
43. Harris M.J., Lambert D.L., Hinkle K.H., Gustafsson B., Eriksson K.: 1987, ApJ 316, 294
44. Harris M.J., Lambert D.L., Smith V.V.: 1988, ApJ 325, 768
45. Hartoog M.R.: 1979, ApJ 231, 161
46. Hartoog M.R., Cowley A.P.: 1979, ApJ 228, 229
47. Herwig F., Blöcker T., Schönberner D., El Eid M.: 1997, A&A 324, L81
48. Hogan 1995: ApJ 441, L17
49. Iben I.Jr.: 1981, ApJ 246, 278
50. Jorissen A., Smith V.V., Lambert D.L.: 1992, A&A 261, 164
51. Kahane C., Cernicharo J., Gomez-Gonzales J., Guelin M.: 1992, A&A 256, 235
52. Kahane C., Gomez-Gonzales J., Cernicharo J., Guelin M.: 1988, A&A 190, 167
53. Kozasa T., Dorschner J., Henning T., Stognienko R.: 1996, A&A 307, 551
54. Lambert D.L., Gustafsson B., Eriksson K., Hinkle K.H.: 1986, ApJS 62, 373
55. Marigo P., Bressan A., Chiosi C.: 1996, A&A 313, 545
56. [9] Mowlavi N.: 1995, Ph.D. thesis, unpublished
57. [9] Mowlavi N.: 1998, in 'Tours Symposium on Nuclear Physics III', ed. H. Utsunomiya, AIP, in press
58. [9] Mowlavi N.: 1998, in 'Abundance profiles: Diagnostic Tools for Galaxy History',

eds. Friedli D., Edmunds M.G., Robert C., Drissen L., ASP Conf. Ser., in press
59. [9] Mowlavi N.: 1998, in 'Pacific RIM Conference on Stellar Astrophysics', eds. Chan K.L., Cheng K.S., Singh H.P., PASP Conf. Proc., in press
60. [9] Mowlavi N., Jorissen A., Arnould M: 1996, A&A 311, 803
61. [9] Mowlavi N., Jorissen A., Arnould M: 1998, A&A, submitted
62. [9] Mowlavi N., Meynet G., this volume
63. Nittler L.R.: 1997, in 'Astrophysical Implications of the Laboratory Study of Presolar Materials', eds. Bernatowicz, Zinner E., AIP, New York, p 59
64. Nittler L.R., Alexander C.M., Gao X, Walker R.M., Zinner E.: 1997, ApJ 483, 475
65. Ohnaka K., Tsuji T.: 1996, A&A 310, 933
66. Olive K.A., Schramm D.N., Scully S.T., Truran J.W.: 1997, ApJ 479, 752
67. Prantzos N.: 1996, A&A 310, 106
68. Renzini A., Voli M.: 1981, A&A 94, 175
69. Rood R., Bania T.M., Wilson T.L.: 1984, ApJ 280, 629
70. Rood R., Bania T.M., Wilson T.L.: 1992, Nature 355, 618
71. Rood R., Bania T.M., Balser D.S., Wilson T.L.: 1997, to appear in 'International Light Elements Workshop', Bern
72. Reeves H.: 1974, ARAA 12, 437
73. Rogers F.J., Swenson F.J., Iglesias C.A.: 1996, ApJ 456, 902
74. Sackmann I.-J., Boothroyd A.I., in 'Evolution os Stars: The Photospheric Abundance Connection' (IAU Symp. 145), eds. Michaud G, and Tutukov A., Kluwer, Dordrecht, p 275
75. Sargent W.L.W., Jugaku J.: 1961, ApJ 134, 777
76. Schaller G., Schaerer D., Meynet G., Maeder A.: 1992, A&AS 96, 269
77. Sweigart A.V., Mengel J.G.: 1979, ApJ 229, 624
78. Smith V.V.: 1989, in 'Cosmic Abundances of matter', ed. Waddington C.J., AIP Conf. Proc. 183, p 200
79. Smith V.V., Lambert D.L.: 1985, ApJ 294, 326
80. Smith V.V., Lambert D.L.: 1985, ApJS 72, 387
81. Smith G.H., Shetrone M.D., Bell R.A., Churchill C.W., Briley M.M.: 1996, AJ 112, 1511
82. Smith G.H., Tout C.A.: 1992, MNRAS 256, 449
83. Smith V.V.,Suntzeff N.B.: 1989, AJ 97, 1699
84. Sneden C., Pilachowski C.A., Vandenberg D.A.: 1986, ApJ 311, 826
85. Straniero O., Chieffi A., Limongi M., Busso M., Gallino R., Arlandini C.: 1997, ApJ 478, 332
86. VandenBerg D.A.: 1992, ApJ 391, 685
87. van den Hoek L.B., Groenewegen M.A.T.: 1997, A&AS 123, 305
88. Vassiliadis E., Wood P.R.: 1993, ApJ 413, 641
89. Wasserburg G.J., Boothroyd A.I., Sackmann I.-J.: 1995, ApJ 447, L37
90. Weiss A., Wagenhuber J., Denissenkov P.A.: 1996, A&A 313, 581
91. Zinner E.: 1994, Meteoritics 29, 555

FLUORINE PRODUCTION FROM AGB AND WR STARS

NAMI MOWLAVI AND GEORGES MEYNET
Geneva Observatory
CH-1290 Sauverny, Switzerland

Abstract. The role of Wolf-Rayet and asymptotic giant branch stars in the production of the Solar System fluorine abundance is analyzed. It is shown that both these stars can be important sources of galactic fluorine. However, the uncertainties affecting the predictions, especially those from asymptotic giant branch stars, do not allow to give quantitative figures yet.

1. Introduction

The origin of the Solar System fluorine was unknown until ten years ago. In 1989, [2] suggested that this element can be produced in helium-burning environment in the presence of neutrons. The nucleosynthesis path follows $^{14}N\,(\alpha,\gamma)\,^{18}F\,(\beta^+)\,^{18}O\,(p,\alpha)\,^{15}N\ (\alpha,\gamma)\,^{19}F$. The required protons would be provided by $^{14}N\,(n,p)\,^{14}C$, with neutrons liberated from $^{13}C\,(\alpha,n)\,^{16}O$. Both ^{13}C and ^{14}N can result from the ashes of H-burning by the CNO cycles.

Two stellar sites have been proposed as candidates for the fluorine nucleosynthesis in He-burning environment. The first one is asymptotic giant branch (AGB) stars. The ^{19}F production in those stars is strongly suggested by observations of fluorine overabundances in red giant S and C stars as compared to normal giants ([4]). Current stellar evolutionary models confirm the operation of the F-nucleosynthesis path in those stars, but fail to reproduce the observed overabundances ([1, 9]; see also [8] in this volume). A source of primary ^{13}C is required, which is difficult to obtain in AGB models (but see [3]). Predictions of the fluorine yield from AGB stars suffer thus from the shortcomings affecting AGB models (see [8] in this volume).

The second stellar site suggested for fluorine production in He-burning environments is Wolf-Rayet (WR) stars ([6, 7]). Contrary to the AGB stars, no observation of fluorine in WR stars has been reported or even attempted yet. However, the nucleosynthesis of that element in those massive stars

K. Nomoto and J.W. Truran (eds.), Cosmic Chemical Evolution, 71–73.

rests on the evolution of WR stars which are relatively easy to model. The fluorine yields from WR stars are thus subject to much less uncertainties than those from AGB stars.

The role of WR stars in the galactic evolution of fluorine, and in particular in the production of its Solar System abundance is recalled in Sect. 2. That of AGB stars is analyzed in Sect. 3.

2. Galactic fluorine from WR stars

While no observation of the F abundance at the surface of WR stars has been reported so far, massive star model predictions for that element can be used to estimate the contribution of WR stars to the galactic fluorine. This has been done by [7], the details of which are not repeated here. Let us recall that these authors arrive to the conclusion that WR stars alone can account for the solar ^{19}F.

3. Galactic fluorine from AGB stars

Let us get an estimation of the ^{19}F contamination by AGB stars to the Solar System fluorine abundance. For that purpose, we consider here that only AGB stars affect the galactic ^{19}F budget, all other possible production or destruction sites being neglected. We make use of a very simple galactic chemical evolution model with the approximations of a closed box and instantaneous recycling. The procedure and notations are described in [7].

Because no fluorine yields are provided by AGB models, we evaluate them from the observed overabundances. If we consider a mean fluorine overabundance of $[^{19}\mathrm{F}/^{16}\mathrm{O}] \equiv [\log {}^{19}\mathrm{F}/^{16}\mathrm{O}] - [\log {}^{19}\mathrm{F}/^{16}\mathrm{O}]_\odot$ in AGB stars, independent of stellar mass and metallicity (no observation of the F abundance in low-metallicity stars is available yet), we have

$$X^s_{19} - X^0_{19} = X^0_{19}\left(10^{[^{19}\mathrm{F}/^{16}\mathrm{O}]} - 1\right) \qquad (1)$$

where X^0_{19} is the initial stellar fluorine mass fraction, and X^s_{19} that at the surface of AGB stars.

Assuming, for simplification, that all AGB stars end as 0.6 $\mathrm{M}_\odot$ white dwarfs, the yields from AGB stars of given initial mass M_i are (Eq. 1 of [7])

$$p^{AGB}_{19}(M_i) = \frac{M_i - 0.6}{M_i}\, X^0_{19}\left(10^{[^{19}\mathrm{F}/^{16}\mathrm{O}]} - 1\right). \qquad (2)$$

In order to derive the net ^{19}F yield y_{19} of a stellar generation, we use an initial mass function $\phi(M_i) = M_i^{-2.7}/\psi$ (the exponent being taken from [5]) with a star formation rate $\psi = 10\ \mathrm{M}_\odot \mathrm{pc}^{-2}\mathrm{Gyr}^{-1}$, and consider that only

AGB stars with $1 \leq M_i \leq 4$ contribute to the interstellar ^{19}F enrichment (stars with $M < 1$ eject their envelope before entering the thermal-pulse AGB phase, while hot bottom burning would prevent AGB stars with $M > 4$ to be efficient fluorine producers). This leads to

$$y_{19} = \frac{1}{1-R} \int_1^4 M_i \, p_{19}^{AGB}(M_i) \, \phi(M_i) \, \mathrm{d}M_i \quad (3)$$

$$\simeq \frac{0.05}{1-R} \, X_{19}^0 \left(10^{[^{19}\mathrm{F}/^{16}\mathrm{O}]} - 1\right) \quad (4)$$

where R is the 'returned fraction' (see [10]).
Taking $R \simeq 0.17$ ([10]) and $[^{19}\mathrm{F}/^{16}\mathrm{O}] \sim 0.5$, the ^{19}F mass fraction $X_{19,\odot}$ at the time $t_\odot$ of Solar System formation is given by

$$X_{19,\odot} = y_{19} \ln[\frac{1}{\sigma(t_\odot)}] \quad (5)$$

$$\simeq 0.25 \, X_{19}^0. \quad (6)$$

AGB stars would thus contribute to as much as 25% of the Solar System fluorine. This is only a *very rough estimate*, the most criticable assumptions being instantaneous recycling and the independency of fluorine production with stellar metallicity. It does show, however, that AGB stars are potentially important contributors to the Galactic fluorine.

References

1. Forestini M., Goriely S., Jorissen A., Arnould M.: 1992, A&A 261, 157
2. Goriely S., Jorissen A., Arnould M.: 1989, in Proc. 5th Workshop in Nuclear Astrophysics; MPA/P1, eds. W. Hillebrandt and E. Müller, Garching, Max Planck Institut für Astrophysik, p. 60
3. Herwig F., Blöcker T., Xchönberner D., El Eid M.: 1997, A&A 324, L81
4. Jorissen A., Smith V.V., Lambert D.L.: 1992, A&A 261, 164
5. Kroupa P., Tout C., Gilmore G., 1993, MNRAS 262, 545
6. Meynet G., Arnould M.: 1993, in Int. Symp. on Nuclear Astrophysics, Karlsruhe, IOP Publishing Ltd
7. Meynet G., Arnould M.: 1996, in *32nd Liège Int. Astroph. Coll.*, eds. Noels A., Fraipont-Caro D., Gabriel M., Grevesse N., Demarque P., pp 89-95
8. [1] Mowlavi N., this volume
9. [1] Mowlavi N., Jorissen A., Arnould M: 1996, A&A 311, 803
10. Tinsley B.: 1980, Fundamentals of Cosmic Physics 5, 287

[1]postscript files available by anonymous ftp on 'obsftp.unige.ch' in 'pub/mowlavi', or in 'http://obswww.unige.ch/~mowlavi/publications/publications.html

DEUTERIUM ABUNDANCE IN THE LOCAL ISM

JEFFREY L. LINSKY AND BRIAN E. WOOD
JILA, Univ. of Colorado and NIST, Boulder, CO 80309-0440 USA

Abstract. Excellent HST/GHRS spectra of interstellar hydrogen and deuterium Lyman-α absorption toward nearby stars allow us to identify systematic errors that have plagued earlier work and to measure accurate values of the D/H ratio in local interstellar gas. Analysis of 12 sightlines through the Local Interstellar Cloud leads to a mean value of D/H = $(1.50 \pm 0.10) \times 10^{-5}$ with all data points lying within $\pm 1\sigma$ of the mean. Deciding whether or not the D/H ratio has different values elsewhere in the Galaxy and beyond will be one of the major objectives of the FUSE mission.

An accurate measurement of $(D/H)_{LISM}$, the D/H abundance ratio in the local interstellar medium (LISM), and an assessment of spatial variations of D/H in the Galaxy are required to address two critically important questions. First, the largest credible D/H ratio in our Galaxy will provide a lower limit to the primordial D/H ratio, $(D/H)_{prim}$, which constrains the critical density of baryons, Ω_B, present in both luminous and "dark" forms. Second, $(D/H)_{LISM}$ is the end result of an incompletely understood complex set of chemical evolution processes in the Galaxy. Comparison of the $(D/H)_{LISM}$ ratio with D/H ratios characteristic of the protosolar nebula and in interstellar gas located elsewhere in the Galactic disk and halo will test our understanding of stellar evolution, stellar mass loss, interstellar physics, and the rate of infall and chemical composition of halo gas.

An accurate value of $(D/H)_{LISM}$ has great importance as careful analyses of absorption spectra toward QSO 1937-1009 and QSO 1009+2956 indicate that $(D/H)_{prim} = (2.4 \pm 0.3) \times 10^{-5}$ (Tytler *et al.* 1996; Tytler & Burles 1996). Thus $(D/H)_{LISM}$ is likely much closer to $(D/H)_{prim}$ than some authors had thought, and our understanding of Galactic chemical evolution will be tested by a measurement of the small difference (perhaps only a factor of 1.5) between $(D/H)_{prim}$ and $(D/H)_{LISM}$.

K. Nomoto and J.W. Truran (eds.), Cosmic Chemical Evolution, 75–79.

Lyman line absorption is generally recognized as the most reliable technique for inferring the present value of D/H in our local environment. This is a consequence of the Sun being surrounded by a cloud of warm, partially ionized gas called the Local Interstellar Cloud (LIC) (e.g., Lallement *et al.* 1995; Wood & Linsky 1997). At the 7,000 K gas temperature in the LIC (Linsky *et al.* 1993; 1995), ionization and grain adsorption is nearly the same for H and D and there are very few molecules. Lyman-α absorption by H and D can be observed with high S/N in HST/GHRS spectra toward nearby stars, provided the hydrogen column density is not so large ($N_{HI} < 10^{18.7}$ cm^{-2}) as to obliterate the D line only 81 km s^{-1}away.

Ferlet *et al.* (1996) reviewed D/H measurements obtained primarily with the Copernicus and IUE spectrographs. These pre-HST studies of Lyman line absorption left a confused picture in which the large range of permitted values of D/H for each LOS is consistent with D/H spatial variations of a factor of 2 or larger on very short spatial scales. The flood of beautiful new GHRS spectra has changed this picture dramatically. The first clear indication of this paradigm shift was the measurement of D/H = $(1.60^{+0.14}_{-0.19})\times10^{-5}$ for the Capella LOS (Linsky *et al.* 1993, 1995). Since this result lies outside of the published error bars for all of the previous results for this LOS, we believe that the older results at least for the late-type stars are unreliable because of systematic errors.

GHRS echelle spectra have far higher S/N and resolution (3.6 km s^{-1}) than Copernicus (15 km s^{-1}) and IUE (25–30 km s^{-1}) spectra. Since the core of the Lyman-α line is very saturated (optical depths of 10^5 – 10^6) and is located on the flat part of the curve of growth, high S/N and spectral resolution are critical for inferring H column densities from the steep outer edges of the core absorption profile. Even with the best available GHRS spectra, however, the column densities are usually more uncertain for H than D. Accurate fits to the outer edges of the core absorption alone may explain much of the previous scatter in the D/H values. We have tried to minimize the following sources of systematic errors:

- GHRS spectra of the Mg II lines and ultra-high resolution spectra of the Na I and Ca II resonance lines typically show many closely spaced narrow velocity components even for short LOS (e.g., Welty *et al.* 1996). We include all known velocity components in our analysis of the H and D lines, but the relative column densities in these components may not be the same as for the metal lines, even for O I and N I which have nearly the same ionization potential as H I.
- In their analysis of the very short (1.3 pc) LOS to α Cen A and α Cen B, Linsky and Wood (1996) found additional absorption on the red side of the Lyman-α line, which they identified as absorption by hot hydrogen in the so-called "hydrogen wall" when the interstellar gas

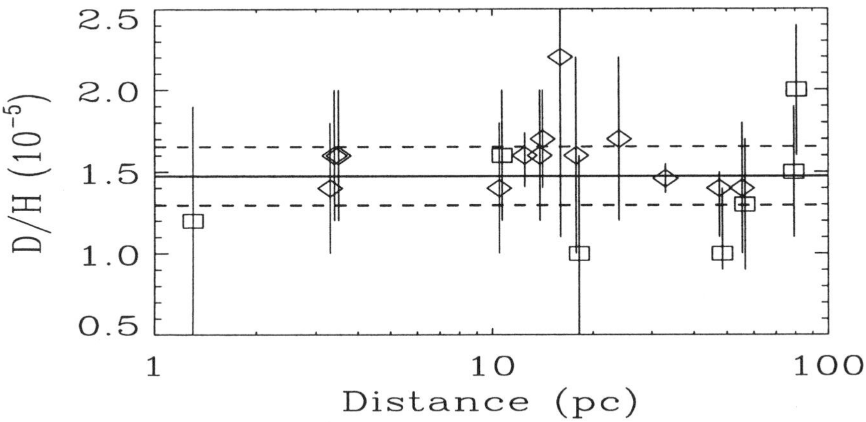

Figure 1. D/H ratios for interstellar gas toward all nearby stars observed with the GHRS. Diamond symbols are for gas in the LIC and square symbols are for other warm clouds. The solid and dashed horizontal lines represent the mean value of D/H and the $\pm 1\sigma$ error of the mean for all data points.

interactes with the solar wind near the heliopause. Subsequent analysis of GHRS spectra of ϵ Ind, ϵ Eri, 61 Cyg A, and 40 Eri A identified additional blue absorption in the Lyman-α line produced by hydrogen walls around these stars (Wood, Alexander, & Linsky 1996; Dring *et al.* 1997; Wood & Linsky 1998). If these additional absorption components were not included, then the inferred H column densities can be a factor of 2 too large and D/H a factor of 2 too small.

- The uncertain stellar Lyman-α emission line profile is the "continuum" against which one measures interstellar absorption. One method for minimizing this uncertainty is to analyze spectroscopic binary systems observed at opposite quadratures (when the orbital radial velocities are a maximum) so that the combined stellar emission line profile has a different shape in the two observations. This allows one to infer the intrinsic stellar emission line profiles and the interstellar absorption nearly independent of each other. Linsky *et al.* (1995) and Piskunov *et al.* (1997) have used this approach to analyze the LOS toward Capella and HR 1099, and Dring *et al.* (1997) have used a similar approach to study the LOS toward σ Gem. High radial velocity stars allow one to derive $N_{\rm H\ I}$ more accurately from the optically thin line wings than from the saturated core (c.f., Wood *et al.* 1996).

Table I lists all of the published D/H values derived for different clouds along the LOS toward nearby stars with GHRS observations. Quantities in parenthesis are assumed rather than derived and are not included in the subsequent analysis. Figure 1 shows the derived D/H ratios for 12 stars with interstellar radial velocities indicating that their LOS pass through the Local Interstellar Cloud (L) and 7 stars with LOS that pass through other nearby warm clouds (G refers to the cloud in the Galactic Center direction, NGP refers to the cloud toward the North Galactic Pole, and numbers indicate how many other clouds are in the LOS). The mean value for the LIC is $(D/H) = (1.50 \pm 0.10) \times 10^{-5}$ and the $\pm 1\sigma$ error bars for all 12 data points are consistent with the mean value. The horizontal and dashed lines in Figure 4 show the mean relation for all data points, $(D/H) = (1.47 \pm 0.18) \times 10^{-5}$. This figure shows that there is no trend in the D/H ratios with distance, and there is no trend with Galactic longitude.

The LIC data indicate that the value of D/H in the tiny region of the Galaxy occupied by the LIC is constant and reasonably well known, but we are just beginning to sample more distant lines of sight which may show different D/H ratios. The data for the other clouds are more scattered with $(D/H) = (1.28 \pm 0.36) \times 10^{-5}$. Whether this scatter represents measurement errors or real D/H differences between the LIC and the other clouds can only be answered with more high quality observations obtained with FUSE and STIS. FUSE, which is scheduled for launch in October 1998, will observe the less opaque high Lyman lines and thus permit us to study D/H further out in the Galactic disk, in the Galactic halo, and toward AGNs.

This work is supported by NASA through grant S-56460-D.

References

Bertin, P., *et al.*: 1995, *A&A*, **302**, 889.
Dring, *et al.*: 1997, *Ap.J.*, in press.
Ferlet, R. *et al.* in *Science with the HST-II*, ed. B. Benvenuti *et al.* (Space Telescope Science Institute, Baltimore, 1996), p. 450.
Gry, C., *et al.*: 1995, *A&A*, **302**, 497.
Lallement, R., *et al.*: 1994, *A&A*, **286**, 898.
Lallement, R., *et al.*: 1995, *A&A*, **304**, 461.
Landsman, W., Sofia, U.J., & Bergeron, P.: 1996, in *Science with the HST-II*, ed. B. Benvenuti *et al.* (Space Telescope Science Institute, Baltimore), p. 454.
Lemoine, M., *et al.*: 1996, *A&A*, **308**, 601.
Linsky, J. L., *et al.*: 1993, *Ap.J.*, **402**, 694.
Linsky, J. L., *et al.*: 1995, *Ap.J.*, **451**, 335.
Linsky, J. L. & Wood, B. E.: 1996, *Ap.J.*, **463**, 254.
Piskunov, N., *et al.*: 1997, *Ap.J.*, **474**, 315.
Tytler, D., Fan, X.-M., & Burles, S.: 1996, *Nature*, **381**, 207.
Tytler, D. & Burles, S.: 1996, in *Origin of Matter and Evolution of Galaxies*, eds. T. Kajino, Y. Yoshii, & S. Kubono (World Scientific Publ. Co.)
Welty, D. *et al.*: 1996, *Ap.J. Suppl.* **106**, 533.
Wood, B.E. & Linsky, J.L.: 1997, *Ap.J*, **474**, L39.
Wood, B.E. & Linsky, J.L.: 1998, *Ap.J*, to appear 10 Jan.
Wood, B. E., Alexander, W. R., & Linsky, J. L.: 1996, *Ap.J.*, **470**, 1157.

TABLE 1. Summary of GHRS Observations of the LISM.

Star	d (pc)	l (°)	b (°)	Grating[†]	Clouds in LOS	D/H (10^{-5}) (in LIC)	D/H (10^{-5}) (others)
α Cen A* [1,2]	1.3	316	-01	EA, EB	G		1.2 ± 0.7
α Cen B [2]	1.3	316	-01	EA, EB	G		1.2 ± 0.7
Sirius [1,3,4]	2.7	227	-09	EB, M	L+1	(1.65)	(1.65)
ϵ Eri [11]	3.3	196	-48	EA, EB	L	1.4 ± 0.4	
ϵ Ind [5]	3.4	336	-48	EA, EB	G/L	1.6 ± 0.4	
Procyon [6]	3.5	214	+13	EB, M	L+1	1.6 ± 0.4	
α Aql [1]	5.0	48	-09	EB	L+2		
α PsA [1]	6.7	21	-65	EB			
Vega [1]	7.5	68	+19	EB	L+2		
β Gem [11]	10.6	192	+23	EA, EB	L+1	1.4 ± 0.4	1.6 ± 0.4
β Leo [1]	12.2	251	+71	EB	L+2		
Capella* [6,7]	12.5	163	+05	EA, EB, M	L	$1.60^{+0.14}_{-0.19}$	
β Cas [8]	14	118	-03	M	L	1.6 ± 0.4	
[11]						1.7 ± 0.3	
β Cet [8]	16	111	-81	EB, M	2	2.2 ± 1.1	
β Pic [1]	16.5	258	-31	EB	L		
α Tri [11]	18	139	-31	EA, EB	L	1.6 ± 0.6	1.0 ± 0.6
λ And [5]	24	110	-15	EA	L	1.7 ± 0.5	
δ Cas [1]	27	127	-02	EB	L		
HR1099* [8]	33	185	-41	EA, EB, M	L+2	1.46 ± 0.09	
G191-B2B [9]	48	156	+07	EB, M	L+2	$1.4^{+0.1}_{-0.3}$	(1.),(1.5)
σ Gem* [11]	56	191	+23	EA, EB	L+1	1.4 ± 0.4	1.3 ± 0.4
HZ 43 [12]	63	054	+84	EA, EB	NGP		(1.6)
31 Com [8]	80	115	+89	EB, M	NGP		1.5 ± 0.4
[11]							2.0 ± 0.4
ϵ CMa [10]	187	240	-11	EB, M	L+5	(1.65)	

Quantities in parenthesis are assumed D/H values that lead to good profile fits.
* These stars were observed twice. Capella, HR 1099 and σ Gem were observed near opposite quadratures.
[†] Gratings: EA = Echelle-A, EB = Echelle-B, M = G140M or G160M.
References: [1] Lallement *et al.* (1995), [2] Linsky & Wood (1996), [3] Lallement *et al.* (1994), [4] Bertin *et al.* (1995), [5] Wood *et al.* (1996), [6] Linsky *et al.* (1995), [7] Linsky *et al.* (1993), [8] Piskunov *et al.* (1997), [9] Lemoine *et al.* (1996), [10] Gry *et al.* (1995), [11] Dring *et al.* (1997), [12] Landsman *et al.* (1996).

ABUNDANCES IN HALO POPULATION STARS

CHRISTOPHER SNEDEN
Department of Astronomy and McDonald Observatory
University of Texas, Austin, TX 78712

1. Introduction

The quest to define early galactic nucleosynthesis has drawn on the efforts of many observers and theoreticians over the past few decades. Significant insights on galactic halo chemical evolution have been revealed through the use of high resolution, high S/N spectroscopy to determine abundances in stars with metallicities ranging over three orders of magnitude. In the process, many galactic nucleosynthesis ideas that were firmly established only a decade ago now are either disproved, or are currently under challenge, or are weighed down with multiple known exceptions. The galactic halo has a rich chemical history, and we are just beginning to understand just how much work still remains before we can decipher its evolution.

In this brief review we will discuss some recent halo abundance studies; for general reviews the reader is referred to, *eg.*, Cayrel (1996) or McWilliam (1997). Our discussion will necessarily be limited. For example, we will not consider abundance trends in LiBeB group, as those elements are explored by other contributors to this volume. The growing literature on light element abundance variations within globular clusters must be left to other reviews. Instead, in the next four sections we discuss in turn the questions of: a) anomalies in α-capture abundance ratios; b) overall trends in "light" and Fe-peak elements in ultra-metal-poor stars; c) relative r- and s-process contributions to neutron-capture elemental abundances as functions of metallicity; and d) new indicators of the very heavy element abundance distribution in the early galaxy.

2. Anomalous α-Capture Abundance Ratios

Pioneering early studies (*eg.*, Wallerstein *et al.* 1962) showed that light even-Z elements O, Mg, Si, Ca and Ti are overabundant in metal-poor halo

K. Nomoto and J.W. Truran (eds.), Cosmic Chemical Evolution, 81–90.

stars: [element/Fe] $\simeq$ +0.3 to +0.5. These elements, generated in α-capture fusion reactions, are efficiently produced in high-mass stars and returned to the ISM in Type II supernovae. Therefore the α-element overabundances in metal-poor stars imply that short-lived high mass stars dominated early galactic halo nucleosynthesis. It has also been asserted that the [α/Fe] ratios of all stars with [Fe/H] metallicities < -1 are nearly identical, thus allegedly proving that the IMF was the same in different parts of the halo or that the halo ISM gas was able to mix on time scales short compared to the formation of subsequent stellar generations.

However, not all halo stars have elevated α-capture abundances; some even have [α/Fe] < 0. Brown *et al.* (1997) found that red giants in two outer halo globular clusters (Rup 106 and Pal 12) have metallicities [Fe/H] $\simeq$ –1.4 and –0.9 but weak α element abundances: $<$[α/Fe]$>$ $\sim$ 0. Carney *et al.* (1997) report that the halo field star BD+80°245 has [Fe/H] $\simeq$ –1.85 but $<$[α/Fe]$>$ $\sim$ –0.25. There is at least one metal-poor star with an anomalously low [Mg/Fe] ratio in the sample of Fuhrmann *et al.* (1995). Finally, Nissen & Schuster (1997) have completed a careful abundance study of 30 stars in the metallicity range $-0.4 >$ [Fe/H] > -1.3; about a quarter of their sample have substantially lower α-element abundances than the rest.

A common characteristic of the low-α stars in these studies is that they have orbits that take them to the outer halo regions of the galaxy. Clearly a new halo sub-population is emerging. The fraction of low-α stars cannot be assessed at the present time, but one can at least speculate on their nucleosynthetic origin. The papers cited above suggest the heavy elements in these stars were enriched in the ejecta of Type Ia supernovae, which are much more efficient producers of iron than the α-elements. Type Ia SNe arise from long-lived stars, and their influence on overall trends in abundance ratios are not apparent until [Fe/H] > -1, when their presence is signaled by declining [α/Fe] ratios. Therefore, Nissen & Schuster suggest that the low-α stars may have formed in ISM regions of abnormally low density, and thus relatively low star-formation rate. This would retard the chemical enrichment of the gas by high-mass Type II SNe, allowing the "delayed" iron production from the Type Ia SNe to occur at relatively low metallicities. Such lower density ISM environments may have been regions of our own galactic halo, or perhaps these stars were produced in nearby dwarf galaxies and later captured by our galaxy; the stellar statistics are insufficient to decide at the present time.

A final clue comes from the depressed levels of the odd-Z light elements sodium and possibly aluminum in the low-α stars. Apparently all such stars discovered to date have [Na/Fe] ≤ -0.3, and this is another indication that Type Ia nucleosynthesis (poor in Na production) was at work. Nissen & Schuster (1997) suggest that searches for these anomalies be made in other halo star samples. Such studies must be done with carefully-selected control samples of normal halo stars, but a small test can be performed here. Pilachowski *et al.* (1996) reported light element abundances for nearly 60 field giants (to compare with the abundances of globular cluster giants). From their sample, we list in Table 1 some mean values of [Na/Fe] in five domains of [α/Fe], where α for a star here is defined as the mean of the Mg, Si, and Ca abundances for that star.

TABLE 1. Field Giants: Mean Abundances

[α/Fe] range	<[Na/Fe]>	No. Stars
+0.05 to +0.14	−0.34±0.07	6
+0.15 to +0.24	−0.28±0.04	10
+0.25 to +0.34	−0.15±0.03	23
+0.35 to +0.44	−0.14±0.05	16
+0.45 to +0.54	+0.04±0.10	4

The correlation between the α-elements and sodium certainly seems present, but confirmation must await a more detailed study. An advantage of Nissen & Schuster's study was the concentration on relatively unevolved nearby stars for which good kinematical information is available. That will not be very easy for the Pilachowski *et al.* (1996) giant stars, most of whom do not have reliable parallaxes even in the post-HIPPARCOS era.

3. Abundance Trends in Ultra-Metal-Poor Stars

Ultra-metal-poor (hereafter, UMP) stars are defined here as those with [Fe/H] < −2.5. This metallicity is the lower limit for globular clusters, and it also is the metallicity at which Fe-peak elements in field stars begin to exhibit severe departures from solar abundance ratios. Much of our knowledge of chemical compositions of UMP stars has come about quite recently, thanks largely to the efforts of the past decade to substantially increase the number of candidate UMP stars. Cayrel (1996) discusses the new UMP surveys at length, but we take special note of the ongoing Beers *et al.* (1992) survey that has significantly augmented the candidate star sample.

Ryan *et al.* (1991,1996) and McWilliam *et al.* (1995) have conducted the most comprehensive abundance surveys of UMP stars; see these papers for references to smaller-sample analyses of UMP stars. The Ryan *et al.* (1996) paper is especially useful, as it includes abundance results from a variety of investigators, thus providing estimates of both intrinsic (scatter) errors within individual studies and external (scale) errors between different studies. Statistical methods are applied to derive robust mean abundance trends, and these are then compared to the galactic chemical evolution models of Timmes *et al.* (1995). The trustworthiness of observational results are also discussed by Ryan *et al.*; for example, the compilation of [Ca/Fe] values in their Figure 2 shows reasonably small scatter (neglecting the few points that may represent α-poor stars; see above) and satisfactory agreement with the Timmes *et al.* predicted trend. By way of contrast, the observational situation for Al is still poor, and Ryan *et al.* give a succinct discussion of problems, which include possible non-LTE effects, clashes between abundances from different Al I transitions employed in studies of high and low metallicity stars, and large offsets between different investigations of the same stars. Their cautions should be heeded; renewed attention should be paid to those elements where observational uncertainties prevent meaningful comparison to theory.

The most noteworthy general result of UMP abundance surveys has been the identification of sharply non-solar abundance ratios among Fe-peak elements, shown most dramatically in small [Cr/Fe] and large [Co/Fe] ratios. Sub-solar [Mn/Fe] in "ordinary" metal-poor stars ([Fe/H] > -2.5) had been found previously (*eg.*, Gratton 1989), but the new surveys identify a further decrease in this abundance ratio at UMP metallicities. The Timmes *et al.* (1995) chemical evolution models predict the trends of [Sc/Fe] and [Mn/Fe] with [Fe/H], and even the possible increase in observed [Ni/Fe] values at [Fe/H] < -3 is suggested by these models. But the Co and Cr variations are much more severe than those suggested by the models, even though there are hints (see Figure 4 of Ryan *et al.* 1996) that some tinkering with model parameters could produce rough agreement. The simplest explanation for the Fe-peak trends is that Fe-peak nucleosynthesis in low-metallicity Type II SNe is driven toward the heavier elements (Co, Ni) at the expense of the lighter ones (Cr, Mn) by invoking an α-rich freezeout to the conditions during the explosive Si-burning.

Left in limbo by this pleasing scenario are Cu and Zn, elements at the high-Z end of the Fe-peak. Extant observations (Sneden *et al.* 1991) suggest that [Zn/Fe] $\simeq 0$ at all metallicities, and that Cu abundances decline sharply with decreasing metallicity ([Cu/Fe] ~ -1 at [Fe/H] ~ -3). Explosive Si-burning is apparently responsible for Zn production, and thus a puzzle emerges: if α-rich freezeout created overabundances of the heaviest Fe-peak elements in UMP stars, why is [Zn/Fe] $\simeq 0$ observed at all metallicities? A small overabundance of Ni should be accompanied by a very large overabundance of Zn. The nucleosynthetic origin of Cu is not at all clear; it may be synthesized in explosive neon burning or it maybe the lightest of the s-process neutron-capture elements. If it could be conclusively shown that Fe-peak element production bypasses Cu, then it might be an effective surrogate for discussions of s-process nucleosynthesis in low metallicity environments. However, the Sneden *et al.* survey included only one star with [Fe/H] < -2.7, and the other UMP surveys do not report abundances for either Cu or Zn. Thus the abundances for these two key elements must be considered simply as unknown in UMP stars, and new observational studies of them should be undertaken.

4. Neutron-Capture Element Abundances at Low Metallicities

Nearly two-thirds of the stable elements in the periodic table are synthesized in neutron bombardment fusion chains. The neutron-capture rates compete with β-decay rates in building almost all nuclei with Z > 30. Low neutron fluxes (the so-called s-process) create unstable nuclei that undergo β decays between successive neutron captures, and element synthesis proceeds along the valley of β-stability. On the other hand, high neutron fluxes (the r-process) create extremely heavy isotopes of nuclei (up to the neutron-drip line) that then rapidly decay back toward the valley of β-stability after the cessation of the neutron flux. The s-process may occur during the quiescent stages of stellar evolution (such as during helium burning), while the most likely site for the r-process is during formation of a neutron star in a Type II supernova detonation (*eg.* Meyer 1994). Thus in the early galaxy when Type II supernovae dominated the light element synthesis (the [α/Fe] ratios of most metal-poor stars are large), one expects substantial r-process contributions to the neutron-capture element abundances as well.

Among the heavier neutron-capture elements, Ba isotopes are synthesized most easily by the s-process, while Eu isotopes are readily made in the r-process. Thus their *elemental* abundances as functions of metallicity can give some indication of the relative r/s contributions in the early galaxy. In Figure 1 we have gathered abundances from several several large-sample studies, including some preliminary results from the new field star abundance survey of Burris *et al.* (1998). Excluded from Figure 1 are abundances of CH stars and other stars whose neutron capture abundance patterns clearly indicate that they have been polluted during their lives by s-process material from within or from a close companion. For most stars [Ba/Fe] values decline at metallicities [Fe/H] < -2, while [Eu/Fe] values remain large. At [Fe/H] ~ -3, [Ba/Eu] ~ -0.8, which is close to the solar-system r-process fractional abundance ratio of these elements (*eg.* Käppeler *et al.* 1989). This point has been made before (see the references in McWilliam 1997); but it is worth noting that all investigators agree on the basic abundance trends. This is very consistent with r-process synthesis from supernovae in the early

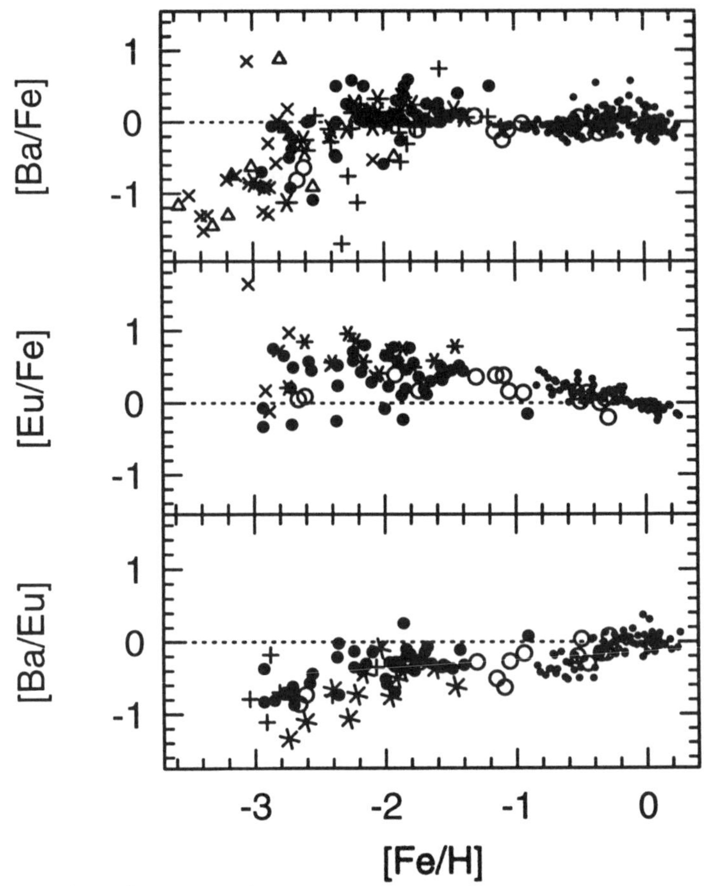

Figure 1. **Ba and Eu abundances as functions of metallicity. The horizontal dotted lines represent the solar abundance ratios. The small circles are from Edvardsson *et al.* (1993) and Woolf *et al.* (1995), the large circles are the subgiant and giant stars of Burris *et al.*** (1998), the asterisks are from Magain (1989) and Zhao & Magain (1990), the open triangles are from Ryan *et al.* (1991,1996), the x's are from McWilliam *et al.* (1995), the plus signs are from François (1996), and the open circles are from Gratton & Sneden (1994).

galaxy, without substantial s-process contribution.

Support for the elemental abundance results comes from a new Ba *isotopic* abundance study by Gacquer & François (1998). They discuss their work in this volume, showing that the width of the Ba II 4554 Å resonance line in the UMP subgiant HD 140283 can be matched only through a higher contribution by the r-process to the Ba abundance in this star than in the solar system. Their analysis clears up the puzzling initial isotopic Ba estimate for this star (Magain 1995), which suggested a large s-process contribution. Ba isotopic analyses should be extended to stars of a large metallicity range, hopefully to settle this question once and for all.

Although the overall trends of [Ba/Fe] and [Eu/Fe] with metallicity are clear, these abundance ratios exhibit $\sim$ 1 dex of star-to-star scatter at any [Fe/H] < -1. However, as argued by Gilroy *et al.* (1988) and extended by Burris *et al.* (1998), this scatter is largely real, and is not dominated by observational errors. Examples abound of metal-poor stars with nearly identical absorption-line strengths for all spectral features except those of the neutron-capture elements. The star-to-star scatter for [Ba/Eu] is substantially smaller than for either [Ba/Fe] or [Eu/Fe]. The [Ba/Eu] ratio is an excellent indicator of the steadily increasing significance of r-process nucleosynthesis with decreasing metallicity, but the ratio of either element with respect to Fe clearly shows the non-homogeneity of "bulk" neutron-capture element abundance levels in galactic halo gas.

This inhomogeneity is vividly illustrated by CS 22892-052, a UMP star from the Beers *et al.* (1992) survey that was included in the McWilliam *et al.* (1995) abundance study. Preliminary reconnaissance of those spectra showed that CS 22892-052 has very large overabundances of all neutron-capture elements (in Figure 1, this star appears at [Fe/H] ~ -3.1, [Eu/Fe] $\sim +1.6$!). Analysis of a new high S/N spectrum by Sneden *et al.* (1996) yielded abundances for 20 neutron-capture elements. The abundances of 15 elements from Z = 56 (Ba) to 72 (Hf) match nearly perfectly a scaled solar-system r-process abundance pattern. This result, combined with the extreme overabundances of all these elements, suggests that *a single* prior (local) supernova explosion seeded the neutron-capture elements in the halo ISM gas cloud that condensed to become CS 22892-052. The early galactic halo should have been chemically inhomogeneous as the first wave of supernovae were occurring, and observational confirmation now exists in the exotic abundances of CS 22892-052 and in the general variations of neutron-capture element levels in UMP stars.

5. New Approaches to the Very Heavy Element Distributions

Thorium (Z = 90) is produced exclusively in the r-process and is unstable, decaying with a half-life of 14.0 Gyrs. Thus its abundance relative to lighter neutron-capture elements can in principle yield an estimate for the "age" of the galaxy. Notable attempts to analyze the Th II feature include those of Butcher (1987), Morell *et al.* (1992), and François *et al.* (1993), who determined either [Th/Nd] or [Th/Eu] ratios. CS 22892-052 has a strong Th II 4019 Å absorption feature, and Sneden *et al.* (1996) used the Th abundance to estimate that the age of neutron-capture material gathered into CS 22892-052 is 15.2 $\pm$ 3.7 Gyrs. Thorium cosmochronometry is no simple task, and Cowan *et al.* (1997 and references therein) discuss the many complications. Briefly, one needs to worry about: [a] line blending problems for the Th II feature (they are severe); [b] the lack, so far, of other Th II features for analysis; [c] the choice of comparison element (the nearly pure r-process element Eu is a happier choice than Nd, which has mixed r- and s-process parentage in solar system material); [d] the production ratio of Th to the lighter neutron-capture elements (so far, a solar-system production ratio has been assumed for lack of a more reliable estimate); and [e] galactic chemical evolution effects on stable and radioactive elements.

All of these points need much work in order for thorium cosmochronometry to be reliable. Progress has been made on the production ratio of Th to other neutron-capture elements. In most metal-poor stars, one has at best measured values of [Th/Eu], and François *et al.* (1993) caution against over interpretation of these values. The *initial* [Th/Eu] ratios generated in early galactic supernovae must be *assumed*, usually by analogy to solar system abundance ratios, in order to use the observed ratios in metal-poor stars to estimate the age of the galactic halo. But there is a gap of 26 elements between Eu and Th! Even in the most favorable case of CS 22892-052, the Z gap is almost 20 between Hf and Th (the Sneden *et al.* 1996 Os abundance is not well determined). Therefore Cowan *et al.* (1996) and Sneden *et al.* (1998) have obtained HST high resolution UV spectra of Os ($Z = 76$), Pt ($Z = 78$) and Pb ($Z = 82$) in three bright very metal-poor giants (CS 22892-052 is too faint for this work). Their analyses demonstrate that these extremely heavy stable elements fit very well into the scaled solar-system r-process distribution of the lighter elements Ba through Eu in these stars. Thus for the r-process components of the neutron-capture elements, the supernovae that produce them seem to have a narrow range of parameters, so that approximately the same abundance distribution of these elements has been generated throughout the galaxy's history. Worries about radical departures from solar system abundance distributions in the r-process material from early galactic supernovae appear to be unfounded, and one may make more confident assumptions about initial Th production for metal-poor stars.

The neutron-capture elements from $31 \leq Z \leq 55$ have been ignored here. The easily observed Sr-Y-Zr element group does not have a clean match to scaled solar system r- or s-process abundance distributions (see, *eg.* Cowan *et al.* 1995). In addition to confusions over the relative r- and s-process contributions to these lighter elements in metal-poor stars, Wasserburg *et al.* (1996) have proposed that there are two distinct r-process production sites that make the lighter and heavier neutron-capture elements (the existence of two s-process synthesis components has been known for some time). If true, it will be difficult to ever understand the nucleosynthetic origin(s) of these elements. To gain insight on this question, Crawford *et al.* (1998) have used Keck spectra of several metal-poor stars to search for other neutron-capture elements with Z $\leq$ 55. In Figure 2 we show their detections of one of the two neutral resonance lines of Ag (Z = 47) and a sample match between synthetic and observed spectra for one of the stars. In this and other stars, it seems that [Ag/Fe] $\sim$ 0, but unfortunately the most metal-poor star of the Crawford *et al.* sample has [Fe/H] $\sim$ −2.2, well above the UMP star domain where the r-process dominates the neutron-capture element abundances. Clearly we need further work on this and other lighter neutron-capture element transitions in the UV spectra of more metal-poor stars. Some of these elements will be accessible to ground-based spectroscopy, but other must be attacked with HST.

We thank John Cowan, Jim Truran, Debra Burris, Caty Pilachowski, Andy McWilliam, Patrick François, and William Gacquer for helpful discussions. This work has been supported by the National Science Foundation, most recently through grant AST-9618364.

References

Beers, T. C., Preston, G. W., & Shectman, S. A. 1992, *AJ*, **103**, 1987
Brown, J. A., Wallerstein, G., and Zucker, D. 1997, *AJ*, **114**, 180
Burris, D. L., Pilachowski, C. A., Armandroff, T., Sneden, C., and Cowan, J. J. 1998, *in preparation*
Butcher, H. R. 1987, *Nature*, **328**, 127
Carney, B. W., Wright, J. S., Sneden, C., Laird, J. B., and Aguilar, L. A. 1997, *AJ*, **114**, 363
Cayrel, R. 1996, *A&A Rev.*, **7**, 217
Cowan, J. J., Burris, D. L., Sneden, C., Preston, G. W., and McWilliam, A. 1995, *ApJ*, **439**, L51
Cowan, J. J., Sneden, C., Truran, J. W., and Burris, D. L. 1996, *ApJ*, **460**, L115
Cowan, J. J., McWilliam, A., Sneden, C., Burris, D. L., and Preston, G. W. 1997, *ApJ*, **480**, 246
Crawford, J., Sneden, C., King, J. R., Deliyannis, C., and Boesgaard, A. M. 1998, *in preparation*
Edvardsson, B., Andersen, J., Gustafsson, B., Lambert, D. L., Nissen, P. E., and Tomkin, J. 1993, *A&A*, **275**, 101
François, P. 1996, *A&A*, **313**, 229
François, P., Spite, F., and Spite, M. 1993, *A&A*, **274**, 821
Fuhrmann, K., Axer, M., and Gehren, T. 1995, *A&A*, **301**, 492
Gacquer, W., and François, P. 1998, *preprint*
Gilroy, K. K., Sneden, C., Pilachowski, C. A., and Cowan, J. J. 1988, *ApJ*, **327**, 298
Gratton, R. G. 1989, *A&A*, **208**, 171
Gratton, R. G., and Sneden, C. 1994, *A&A*, **287**, 927
Käppeler, F., Beer, H., & Wisshak, K. 1989, *Rep. Prog. Phys.*, **52**, 945
Magain, P. 1989, *A&A*, **209**, 211
Magain, P. 1995, *A&A*, **297**, 686
McWilliam, A. 1997, *ARA&*, in press
McWilliam, A., Preston, G. W., Sneden, C., & Searle, L. 1995, *AJ*, **109**, 2757

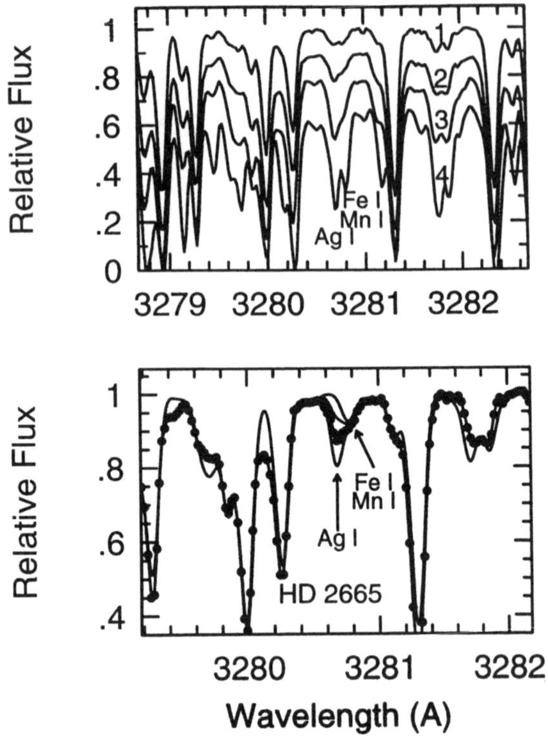

Figure 2. **The detection of silver in metal-poor stars (Crawford *et al.* 1998). In the upper** panel we show the blended Ag I line in BD +37 1458 (star 1), HD 2665 (2), HD 6755 (3), and HD 103095 (4). In the bottom panel the observed HD 2665 spectrum is given by the solid line with filled circles representing the individual spectrum points, and syntheses for this low metallicity star ([Fe/H] $\simeq$ −2) are shown for [Ag/Fe] = −12.0 (essentially no Ag), −0.5, −0.1 (the best fit; buried under the data points), and +0.2.

Meyer, B. S. 1994, *ARA&A*, **32**, 153
Morell, O., Källander, D., & Butcher, H. R. 1992, *A&A*, **259**, 543
Nissen, P. E., and Schuster, W. J. 1997, *A&A*, **326**, 751
Pilachowski, C. A., Sneden, C., and Kraft, R. P. 1996, *AJ*, **111**, 1689
Ryan, S. G., Norris, J. E., and Bessell, M. S. 1991, *AJ*, **102**, 303
Ryan, S. G., Norris, J. E., and Beers, T. C. 1996, *ApJ*, **471**, 254
Sneden, C., Gratton, R. G., and Crocker, D. A. 1991, *A&A*, **246**, 354
Sneden, C., McWilliam, A., Preston, G. W., Cowan, J. J, and Burris, D. L. 1996, *ApJ*, **327**, 298
Timmes, F. X., Woosley, S. E., and Weaver, T. A. 1995, *ApJS*, **98**, 617
Wallerstein, G., Greenstein, J. L., Parker, R., Helfer, H. L., and Aller, L. H. 1963, *ApJ*, **137**, 280
Wasserburg, G. J., Busso, M. & Gallino, R. 1996, *ApJ*, **466**, L109
Woolf, V. M., Tomkin, J., and Lambert, D. L. 1995, *ApJ*, **453**, 660
Zhao, G., and Magain, P. 1990, *A&A*, **238**, 242

ABUNDANCES IN THE GALACTIC DISK

BENGT EDVARDSSON
Uppsala Astronomical Observatory
Box 515, S-751 20 Uppsala, SWEDEN

1. Introduction

The empirical study of the build-up of chemical abundances in the galactic disk is an important route to our understanding of the history of the Galaxy. It supplies a wealth of constraints to the models and simulations being done. There are, however, still many unanswered questions concerning the details of the abundance patterns, and many uncertainties and possibly inconsistencies in the data at hand. Ideally one would like to map the abundances of individual elements as a function of time and 3-dimensional position in the disk. The study of the ISM today and of the surfaces of individual stars as probes of the ISM from the time and position of their formation help us in this endeavour.

Here I will constrain myself to discussion of abundances and abundance gradients in the local galactic disk and continue with the scatter in these abundances, presently and historically.

2. ISM Abundances

Abundances in the current (age less than 1 Gyr) disk interstellar medium can be obtained from observations of e.g. compact H II regions, field B main sequence stars, stars in young open clusters, absorption lines in the ISM, and young planetary nebulae. The different methods do not always give the same results for objects with similar birth places, which may be due to real ISM inhomogeneities on not too large scales.

For a long time, discussions of oxygen abundances derived from B main-sequence stars have indicated that there is no or only a very weak radial abundance gradient in the galactic disk. H II regions and young planetary nebulae, on the other hand, have fairly consistently suggested an oxygen gradient of about -0.06 to -0.09 dex/kpc (lower abundances at larger

K. Nomoto and J.W. Truran (eds.), Cosmic Chemical Evolution, 91–96.

galactocentric distances), cf. e.g. Afflerbach et al. (1997) as well as typically sub-solar O/H abundance ratios in the solar neighbourhood Costa et al. (1996). This inconsistency seems to have changed recently with the extensive analysis of B stars in the field and in open clusters by Smartt & Rolleston (1997). Their new abundances, combined with distance redeterminations, now agree very well with the "nebular" results, both as regards the absolute scale and the gradient of -0.07 ± 0.01 dex/kpc for galactocentric distances between 6 and 18 kpc.

Nebular results for C, N, Ne, S, and Ar indicate similar gradients, and are typically 0.2 to 0.3 dex lower than solar in the local disk, cf. e.g. Maciel & Köppen (1994), Esteban & Peimbert (1995), Cardelli et al. (1996), Mathis (1996), Afflerbach et al. (1997).

The low abundance suggested for e.g. carbon causes difficulties for our understanding of the interstellar extinction (Mathis 1996). There is not enough carbon to explain the total amount of extinction. This problem has been dubbed the "C/H crisis" (Kim & Martin 1996) and may also be present for other elements, why modifications are sought to the standard model of ISM dust particles – still, however, without satisfactory results (Dwek 1997).

This pattern of sub-solar abundances may not be parallelled by iron and iron-peak elements. Nissen (1988) presented Strömgren $uvby$-β photometry of solar-type stars in young open clusters. Since the metallicity [M/H] derived from the m_1 index is a direct measure of the line blanketing, which for these stars is predominantly caused by lines of iron and iron-peak elements, it is often interpreted as an iron abundance, [Fe/H][1] Fig. 1a shows the iron abundances in Nissen's 10 youngest clusters relative to the Sun. In Fig. 1b are actual iron abundances derived from solar-type dwarfs in 6 open clusters observed with high-resolution spectroscopy by Boesgaard (1989). The 5 clusters in common between the two investigations are identified by name in the figure. The agreement between the results using completely different techniques is excellent; it is only for Coma that the results differ by 0.055 dex (14%) and the error bars do not meet, which is only what one may expect from a statistical point of view. The typical iron abundance in young open clusters appears to be very close to solar: The 10 youngest clusters of Nissen have a mean metallicity [Fe/H]= $+0.04 \pm 0.02$ (error in the mean), and for the 6 Boesgaard clusters the corresponding figures are [Fe/H]= $+0.01 \pm 0.04$. The mean values of [Fe/H] for the 5 clusters in common between the investigations differ by 0.020 dex. The systematic errors are difficult to judge (a major source may be the still uncertain effective temperature scale), but they appear unlikely to be as large as 0.15 dex and

[1]The bracket notation, [X/Y], indicates logarithmic (number-) abundance ratios relative to the Sun, e.g.: $[\mathrm{Fe/H}] = \log(N_{\mathrm{Fe}}/N_{\mathrm{H}})_{\mathrm{Star}} - \log(N_{\mathrm{Fe}}/N_{\mathrm{H}})_{\odot}$

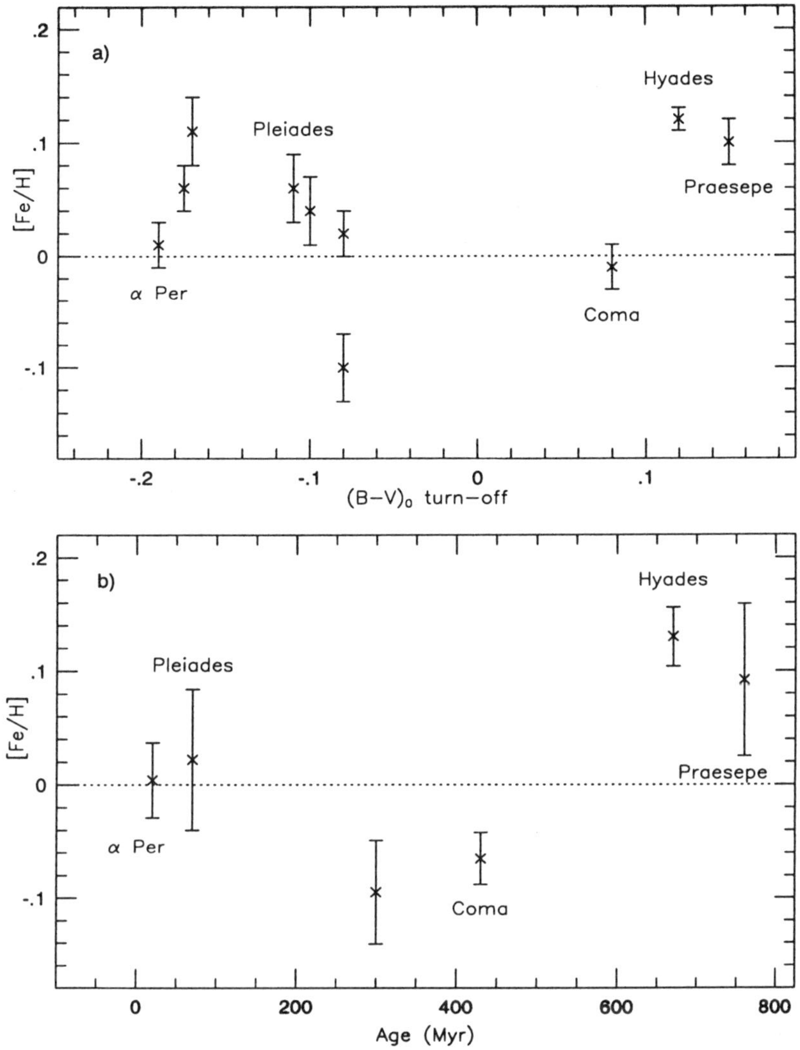

Figure 1. **Iron abundances in nearby, young open clusters. Panel a) shows metallicities (iron abundances) derived from Strömgren photometry by Nissen (1988). Panel b) shows** spectroscopic iron abundances by Boesgaard (1989). Note the very good agreement for the 5 clusters in common between the two very different analyses

are probably less than 0.10 dex considering the similarity of the stars to the relatively well known reference object called the Sun.

Taken at face value, the local ISM appears to be deficient in light elements produced in massive stars with short life times. But it is not deficient in iron, which is predominantly made in Sn Ia with longer-lived intermediate and low-mass progenitors. We don't yet know whether this difference between iron and the lighter elements is real or due to some systematic errors in the analyses. Considering the different objects and methods used to derive these abundances it would therefore be very valuable to "cross confirm" these results. To this end, carbon and oxygen abundances should be derived from G dwarfs with effective temperatures within 100 K of the Sun by means of the weak but reliable [C I] 8727 Å and [O I] 6300 Å lines, which are quite insensitive to uncertainties in effective temperatures. For the B stars, iron abundances may be derived from Fe III lines in the UV as suggested by Smartt & Rolleston (1997).

3. The ISM abundance scatter

The intrinsic scatter in the nebular abundances for objects at a certain galactocentric distance is often difficult to estimate and to disentangle from random errors in the analyses, but is probably on the order of 0.10 dex.

There are, however, young objects for which the abundance scatter is larger than the random errors in the analyses, e.g. the solar-type dwarfs in open clusters studied by Nissen (1988) and Boesgaard (1989) and discussed in the previous section (see Fig. 1). The cluster-to-cluster abundance scatter in Nissen's data is ±0.07 dex, while the relative abundances are accurate to ±0.01 to 0.03 dex. In Boesgard's sample the scatter is ±0.09 dex while the relative error bars range from 0.023 to 0.067 dex. From this I conclude that there may be abundance inhomogeneities of about 0.1 dex on physical scales of less than a kpc in the interstellar medium. It would be very interesting to find out just how small these scales really are. Is the difference mainly between spiral arm and inter-arm regions or could there be substantial abundance gradients within individual star-forming regions?

There is also some evidence that there has always been abundance scatter in the galactic disk. The sample selection in the detailed study of 189 solar-type field stars of Edvardsson et al. (1993) is biased towards low-metallicity stars and against possible old high-metallicity stars, i.e. it is not a complete volume-limited sample and should therefore not be taken as representative for the local disk population and used for studies of age-metallicity relations or the distribution of metallicities in the disk. Nevertheless, we tried to make a rudimentary volume correction, and claim that throughout the existence of the galactic disk there was always an abundance scatter of at least 0.10 to 0.15 dex in [Fe/H] for stars born at the same time and at the same galactocentric distance. An important additional result is that there is no evidence for any scatter in [α/Fe] larger than the error in the relative abundances, about ±0.05 dex (Table 17). The α elements (which are thought to be almost exclusively produced in supernovae of short-lived, massive stars; Types II, Ib and Ic) were represented by a straight mean of elements with α-element "behaviour": $[\alpha/H] = \frac{1}{4}([Mg/H] + [Si/H] + [Ca/H] + [Ti/H])$. This forces the strong conclusion that any scatter of this kind CAN NOT be caused by local pollution of the ISM by e.g. bursts of star formation. While such bursts might cause local enhancements of α elements made in massive stars, they could not cause correlated enhancements of the same magnitude in iron-peak nuclei, since these are to at least 50% produced in Sn Ia with long-lived progenitors. It therefore appears unlikely that "self-pollution" in star-forming regions is an important process in the galactic disk. It was concluded that Larsson's (1972) idea of infall of metal-poor gas on the galactic disk, which would mix with and homogeneously dilute the ISM and trigger star formation is a possible mechanism to cause the scatter. Another possibility which has been suggested by Wielen et al. (1996) is that all the scatter is due to the orbital mixing of stars born at different galactocentric distances, combined with smooth radial abundance gradients in the disk. This may explain the properties of the scatter. The existence of the required orbital scatter is based on the motions of stars in the solar vicinity, but the results of Strömgren (1987) and Sommer-Larsen & Antonuccio-Delogu (1993) contradict the existence of the large velocity scatter required for this explanation. Furthermore, it is by no means clear what objects or processes might cause the required strong disturbances of stellar orbits.

The presently unclear empirical situation concerning the dynamical heating of the galactic disk may soon be clarified by the ongoing work by Nordström et al. (1998). With a homogeneous material of Strömgren *uvby-β* photometry of about 18,000 stars in the solar neighbourhood, and for a sub-sample of about 5,000 stars complete distances and proper motions, accurate radial velocities and ages determined from isochrones will be available. This will constitute an excellent data base for the study of metallicity and metallicity scatter, stellar dynamics and orbital scatter (among other issues), and should finally settle the question concerning heating and mixing of stellar orbits in the disk.

4. Conclusions

We need to confirm whether the lighter elements like C, N, O, Ne, S, Ar really are underabundant in the ISM, while iron has close to the solar abundance. The abundances of carbon and oxygen in the ISM should therefore be measured also in solar-similar dwarfs in young open clusters, while iron abundances could be derived from the UV spectra of main-sequence B stars.

The scales of abundance inhomogeneities in the local ISM should be studied by homogeneous analyses of "identical" objects within individual star-forming regions and between different ones.

References

Afflerbach A., Churchwell E., Werner M.W. 1997, ApJ 478, 190
Boesgaard A.M. 1989, ApJ 336, 798
Cardelli J.A., Meyer D.M., Jura M., Savage B.D. 1996, ApJ 467, 334
Costa R.D.D., de Freitas Pacheco J.A., de Franca J.A.,JR., 1996, A&A 313, 924
Dwek E. 1997, ApJ 484, 779
Edvardsson B., Andersen A., Gustafsson B., Lambert D.L., Nissen P.E., Tomkin J. 1993, A&A 275, 101
Esteban C., Peimbert M. 1995, RMxAA Serie de Conferencias 3, 133
Kim S.-H., Martin P.G. 1996, ApJ 462, 296
Larson R.B. 1972, Nature 236, 21
Maciel W.J., Köppen J. 1994, A&A 282, 436
Mathis J.S. 1996, ApJ 472, 643
Nissen P.E. 1988, A&A 199, 146
Nordström B., Andersen J., Olsen E.H., Mayor M. 1998, in preparation
Smartt S.J., Rolleston W.R. 1997, ApJ 481, L47
Sommer-Larsen J., Antonuccio-Delogu V. 1993, MNRAS 262, 350
Strömgren B. 1987, in "The Galaxy", Proc. NATO Adv. Study Inst., eds. G.Gilmore, R.Carswell, Reidel, Dordrecht, p. 229
Wielen R., Fuchs B., Dettbarn C. 1996, A&A 314, 438

ABUNDANCES IN A HOT HORIZONTAL-BRANCH STAR IN THE OLD OPEN CLUSTER NGC/6791

R.C. PETERSON
Astrophysical Advances, Palo Alto, CA
and UCO/Lick Observatories, Santa Cruz, CA

AND

E.M. GREEN
Steward Observatory, University of Arizona, Tucson, AZ

Abstract.

The open cluster NGC 6791 is more metal-rich than any other open cluster, yet several hot blue horizontal branch (BHB) stars are probable members. We have performed an abundance analysis of the coolest such star, 2-17, which has a radial velocity and a proper motion which support cluster membership. We find parameters typical of a metal-rich BHB star, with rotational velocity = 16 ± 1 km/s, an effective temperature $T_{\rm eff}$ = 7250 ± 150 K, log gravity = 3.5 ±0.5 dex, and an iron abundance more than twice solar, [Fe/H] = +0.4 ± 0.1 dex. The light even-Z elements Mg, Ca, and Si are further enhanced, ruling out A-star peculiarities. Its analysis thus establishes the cluster metallicity accurately, free from serious blending which is problematical in metal-rich giants. Abundances of C, N, and O suggest little mixing. Visual comparison of the spectra of red HB stars in the cluster with that of the field star μ Leo indicates that the spectra are indistinguishable except for stronger CN in the former.

The high iron abundance, the light-element overabundances, the spectral similarity to strong-lined field giants, and the extensive presence of hot stars all highlight the fact that this old open cluster provides an excellent template for the study of metal-rich extragalactic systems. Properties of cluster members should establish not only accurate abundances for comparison with scenarios of nucelosynthesis and star formation, but also provide a testbed for ideas of the production of hot stars in a metal-rich environment, and so for the generation of the ultraviolet upturn in elliptical galaxies.

K. Nomoto and J.W. Truran (eds.), Cosmic Chemical Evolution, 97–102.

1. Introduction

The open cluster NGC 6791 is now considered both the oldest and the most metal-rich known. Its age is 8 – 10 Gyrs, twice as old as the canonical solar-metallicity cluster M67 (Garnavich et al. 1994; Demarque, Green, & Guenther 1992; Tripicco et al. 1995). That its metallicity is significantly above solar is suggested from moderate-resolution spectroscopy and from a mismatch of its color-magnitude diagram (CMD) with solar-metallicity isochrones. Tripicco et al. (1995) find [Fe/H] = +0.27 to +0.44.

The cluster population is rich. In addition to about a dozen red giants and two dozen red horizontal-branch stars, the cluster has several very hot HB stars (Kaluzny & Udalski 1992). Liebert et al. (1994) have shown that the extremely blue stars are mostly sdB/sdO stars and at least 3 or 4 are likely cluster members, the first ever discovered in an open cluster. These may provide the key to the puzzling upturn in ultraviolet flux below 1500Å seen in many high-metallicity systems (Burstein et al. 1988; Ferguson et al. 1991; Rich, Minniti, & Liebert 1993).

Cooler stars are less certain as members because of heavy field contamination. One is an almost certain member, however, with both a radial velocity and a preliminary proper motion (Cudworth 1994) which is consistent with membership. This star, 2-17, has B-V = 0.37 and V = 15.0, placing it at the extreme red end of the blue horizontal branch (BHB). Its analysis was undertaken to confirm its status as an evolved metal-rich BHB star and to establish abundances for the cluster as a whole from a spectrum in which line strengths and line blending are quite similar to that of the Sun, greatly reducing systematic errors.

2. Abundance Analysis

Echelle spectra of 30,000 FWHM resolution were obtained on the 4m telescope of the National Optical Astronomical Observatories at Kitt Peak near Tucson, AZ, with a TEK 2048x2048 CCD detector. Multiple exposures totaling several hours produced a spectrum with S/N = 30 per pixel near 6000Å and covering 4500 – 7800Å. The spectrum was reduced following standard procedures using the package IRAF. After cosmic-ray removal, dark and bias were removed by subtracting coadded sequences of dark and bias calibration frames. Flatfielding was accomplished by pixel-by-pixel division of the coadded spectrum by the coadded sequence of quartz lamp exposures. The wavelength solution was found by a two-dimensional fit to Th-Ar emission features and shifted by 52 km/s to the stellar rest frame.

The abundance analysis proceeded in three stages: spectral synthesis to identify unblended FeI and FeII lines; an equivalent width analysis of these lines to establish stellar temperature $T_{\rm eff}$, gravity $\log g$, and [Fe/H], the logarithmic iron abundance with respect to the Sun; and spectral synthesis of the entire 5000 – 6500Å and 7500 – 7800Å region plus isolated additional sections to derive abundances for additional elements. Models interpolated in the grid of Kurucz (1993) were used with species number densities calculated explicitly for the elemental distribution involved. We employed the Kurucz programs SYNTHE and WIDTH.

For the spectral synthesis, atomic line parameters in the 5000 – 7500Å region were taken directly from the fit to the solar intensity spectrum by Peterson, Dalle Ore, & Kurucz (1993), hereafter PDOK. To verify parameters for lines outside this wavelength region, calculations were also made for the solar intensity spectrum observed by Brault and reduced by Kurucz (see Kurucz 1991).

For the equivalent width analysis, we employed a set of 15 FeI lines and 4 FeII lines in the 6000 – 6500Å region which were unblended and were included in the analysis of metal-poor giants by Kraft et al. (1995). Their equivalent widths in 2-17 ranged from 15 to 112mÅ for FeI and from 60 to 103mÅ for FeII. We ran these lines twice through each model, using first the transition probabilities from Kraft et al. and then those of PDOK, to derive stellar v_t and $T_{\rm eff}$ from the dependence of the deduced iron abundance on equivalent width and lower excitation potential. The results were as follows: for the Kraft et al. transition probabilities, $T_{\rm eff}$ = 7330K, v_t = 2.5 km/s, log(Fe/H) = -3.96, and logg = 4.0; for the PDOK transition probabilities, $T_{\rm eff}$ = 7250, v_t = 2.8 km/s, log(Fe/H) = -3.98, and $\log g = 3.5$. The uncertainty in the mean in both cases is about 80K for $T_{\rm eff}$, 0.5 dex for $\log g$, 0.06 in log(Fe/H), and 0.3 km/s for v_t, with the $T_{\rm eff}$ and v_t uncertainties dictating that of log(Fe/H). The one-sigma deviation of a single line about the mean for FeI and FeII respectively for FeI and FeII respectively was 0.16 and 0.07 for the Kraft et al. values and 0.13 and 0.06 for those from PDOK, which for FeI probably reflect the uncertainties in the transition probabilities themselves. Their systematic uncertainties are difficult to estimate, and seem to dictate the overall uncertainty in [Fe/H]. Kraft et al. adopt log(Fe/H) = −4.48 for the Sun, while PDOK adopt the earlier Grevesse & Anders (1989) value of −4.33; thus the relative iron abundance is [Fe/H] = +0.51 from the Kraft analysis, while [Fe/H] = +0.35 results from the PDOK analysis.

Subsequent spectral synthesis for the star was done using a model with $T_{\rm eff}$ = 7250, $\log g$ = 3.0, v_t = 2.7 km/s, and [Fe/H] = +0.3 and PDOK line parameters. This yielded abundances good to 0.1 dex for several additional species. Mg and Si are enhanced by 0.25 dex with respect to iron and Ca by 0.1 dex, while Na is enhanced by 0.4 dex. The very weak OI triplet at 6156Å indicates [O/Fe] = 0.0, while the the strong 7774Å OI triplet is matched by [O/Fe] = +0.2; the latter is often anomalously strong in stars of such temperature. Three CI lines suggest [C/Fe] = 0.0, while two NI features indicate [N/Fe] = +0.5.

The abundance analysis demonstrates that 2-17 has the properties expected of a metal-rich blue HB star. Its radial velocity agrees to wtihin 2 km/s with that of the cluster mean, and was constant to 1 km/s during the two nights of echelle observations. Its rotational velocity is 16 km/s, typical of blue HB stars in globular clusters (Peterson, Rood, & Crocker 1995). Its overall iron abundance is high. Its relative abundance distribution shows no sign of chemical pecularities typical of Population I A stars; for example, calcium is underabundant in such stars.

The relative abundance distribution suggests that little if any mixing has occurred which might have altered C, N, or O abundances, since [C/Fe] = 0. We cannot rule out the possibility of an initially high [C/Fe] value and subsequent reduction of C and O with enhanced N the result. However, a simpler interpretation is that the star reflects the abundance distribution of products of a type II supernova of solar metallicity or higher, in that it shows to a modest extent the enhancement of light even-Z species which is ubiquitous in halo stars and attributed to this process (Wheeler, Sneden, & Truran 1989). The Na and N enrichment could be the natural consequence of the high metallicity of the supernova.

3. Spectral Comparison of Red Giants

A subsequent run with the echelle spectrograph of Lick Observatory at the 3m telescope on Mt. Hamilton during four nights in July and August 1997 yielded spectra of 35,000 resolution and S/N typically 20 per pixel, for nearly thirty stars whose colors suggested they are red HB members of NGC 6791. Radial velocities confirmed this in almost all cases. Seven such spectra were coadded pixel-by-pixel to produce a preliminary composite spectrum typical of such stars.

The resulting spectrum was overlaid on a high S/N spectrum of similar resolution previously obtained for the field star μ Leo, the prototypical super-metal-rich star. The result was surprising: details of spectral absorption lines matched extremely well. The only discernible difference appeared where CN features dominated; these were somewhat stronger in the composite NGC 6791 spectrum than in μ Leo.

4. Conclusions

In all the above aspects, the cluster NGC 6791 seems to provide a template for the study of metal-rich extragalactic systems. The iron abundance of the cluster is well established by the blue HB star 2-17, and is high, 0.35 – 0.5 dex. The abundance distribution of that star probably reflects that of the cluster to a large extent; it is most simply interpreted as the result of the products of Type II supernova explosions whose progenitors were of solar metallicity or higher. The cluster's RHB stars are at least as strong-lined as the field super-metal-rich prototype μ Leo. Sodium and magnesium are strong features in such stars, as they are in the integrated visible spectra of luminous external galaxies. Moreover, the cluster harbors extremely blue sdB/sdO stars, which could be the cause of the upturn in ultraviolet flux seen in the most luminous ellipticals.

The cluster serves as a caution to a straightforward scenario of heavy-element production within the Milky Way. It is old, but extremely metal-rich, defying an age-metallicity relation. It seems difficult to produce such a high heavy-element abundance in situ, especially if Type II supernovae were involved, since the products would escape the cluster's small potential well. However, if formed by collisions among enriched molecular clouds, or as a merger with an external metal-rich object, it remains difficult to see how the fragile binding energy could be maintained. Given its orbit not far from the disk of the galaxy, its dynamical stability is a general issue.

The cluster's individual members pose similar challenges to stellar evolution and stellar population studies. Production of both sdO/sdB stars and a BHB of intermediate temperature is very difficult with single-star evolution; perhaps binary mechanisms are involved, or the triggering of extreme mass loss by another mechanism. The colors of 2-17 point to the need for broader coverage in surveys of the Galactic disk designed to pick up blue stars. Because of its heavy line blanketing (due to its high metallicity), the visual colors are redder than normal for a stable HB star but rather typical of an RR Lyrae star; as such it might have been missed in many BHB surveys.

5. References

Burstein, D., et al. 1988, ApJ, 328, 440

Cudworth, K. M. 1994, private communication

Demarque, P., Green, E. M., & Guenther, D. B. 1992, AJ, 103, 151

Ferguson, H. C., et al. 1991, ApJ, 382, L69

Friel, E. D., & Janes, K. A. 1991, A&A, 267, 75

Garnavich, P. M., VandenBerg, D. A., Zurek, D., & Hesser, J. E. 1994, AJ, 107, 1097

Grevesse, N., & Anders, E. 1989, in *Cosmic Abundances of Matter*, AIP Conf. 183, 1

Kaluzny, J., & Udalski, A. 1992, A&A, 42, 29

Kraft, R. P., Sneden, C., Langer, G. E., Shetrone, M. D., & Bolte, M. 1995, AJ, 109, 2586

Kurucz, R. L. 1991, Rev. Mex. A. Ap.
Kurucz, R. L. 1993, CDROM series
Liebert, J., Saffer, R. A., & Green, E. M. 1994, AJ, 107, 1408
Peterson, R. C., Dalle Ore, C. D. M., & Kurucz, R. L. 1993, ApJ, 404, 333
Peterson, R. C., Rood, R. T., & Crocker, D. A. 1995, 453, 214
Rich, R. M., Minniti, D., & Liebert, J. 1993, ApJ, 406, 489
Tripicco, M. J., Bell, R. A., Dorman, B., & Hufnagel, B. 1995, AJ, 109, 1697
Wheeler, J. C., Sneden, C., & Truran, J. T. 1989, ARA&A, 26

ON THE PROGENITORS OF TYPE IA SUPERNOVAE

X.-D. LI
Department of Astronomy, Nanjing University
Nanjing 210093, P. R. China

AND

E. P. J. VAN DEN HEUVEL
Astronomical Institute, University of Amsterdam
1098 SJ Amsterdam, The Netherlands

1. Introduction

Supersoft X-ray sources (hereafter SSS) are a class of luminous (bolometric luminosity $\sim 10^{37} - 10^{38}$ erg s^{-1}) objects with a characteristic radiation temperature of 30 to 60 eV (Hasinger 1994; Kahabka & Trümper 1996). The most popular model for SSS is that they are massive white dwarfs steadily burning nuclear fuel accreted from a more massive binary companion at a rate near or above the Eddington limit (van den Heuvel et al. 1992).

Detailed population studies on SSS and their evolutionary connection with type Ia supernovae (SNe Ia) have been made by Rappaport et al. (1994), Yungelson et al. (1996) and Di Stefano & Nelson (1996). Recently, Hachisu, Kato & Nomoto (1996) found a new strong wind solution when the mass accretion rate of a white dwarf exceeds a critical value, thus avoiding the common envelope evolution. Motivated by the work of Hachisu et al. (1996), we have performed evolutionary calculations of white dwarf binaries, to investigate the population of the possible SN Ia progenitors.

2. The model

We consider a C-O white dwarf accreting from a companion star with solar chemical composition (X=0.7, Y=0.28, Z=0.02). For the outcome of the evolution, we only consider the Chandrasekhar-mass SN Ia explosion of the white dwarf with an initial mass less than $1.2M_{\odot}$. The companion star is assumed to be lobe-filling, and mass transfer occurs through Roche-lobe overflow. The system can be specified by three parameters: the initial white dwarf mass $M_{\rm WD,0}$, the initial donor mass $M_{\rm d,0}$ and the orbital period

K. Nomoto and J.W. Truran (eds.), Cosmic Chemical Evolution, 103–108.

$P_{\rm orb,0}$ at the beginning of mass transfer. We have followed the evolution of the binary systems for various sets of these parameters, using an updated version of the evolution code developed by Eggleton (1971).

Whether the white dwarf can grow in mass is determined by the "accumulation ratio" α, the fraction of the accreted hydrogen that converts into elements heavier than helium. The value of α is related to mass loss from the white dwarf during hydrogen and helium burning, i.e., $\alpha \equiv \alpha_{\rm H}\alpha_{\rm He}$, where $\alpha_{\rm H}$ and $\alpha_{\rm He}$ are the fraction of the transferred mass accumulated during hydrogen and helium burning, respectively.

It has been shown (e.g. Nomoto et al. 1979) that steady hydrogen burning on the white dwarf surface can occur when $\dot{M}_{\rm low} \le \dot{M} \le \dot{M}_{\rm up}$, where $\dot{M}$ is the mass transfer rate, $\dot{M}_{\rm up} \simeq 8.5\times 10^{-7}(M_{\rm WD}/M_\odot - 0.52)\, M_\odot\ {\rm yr}^{-1}$ and $\dot{M}_{\rm low} \simeq 0.4\dot{M}_{\rm up}$. As $\dot{M} < \dot{M}_{\rm low}$, hydrogen burning is unstable and occurs in flashes, in which part of the white dwarf envelope may be ejected. For the hydrogen accumulation ratio $\alpha_{\rm H}$, we have taken the following assumptions: (1) if $\dot{M} > \dot{M}_{\rm up}$, we adopt the strong wind solution by Hachisu et al. (1996), which allows burning of hydrogen into helium at a rate limited to $\dot{M}_{\rm up}$, the excess material being blown off in the wind, i.e., $\alpha_{\rm H} = \dot{M}_{\rm up}/\dot{M}$; (2) if $\dot{M} \le \dot{M}_{\rm up}$, we fit the data of the accumulation ratio calculated by Prialnik & Kovetz (1995) for hydrogen burning and flashes (for each combination of $M_{\rm WD}$ and $\dot{M}$) with the following conditions: (i) when $\dot{M}$ declines below $3\times 10^{-8} M_\odot\ {\rm yr}^{-1}$, a strong nova explosion is assumed to occur and no mass accumulates on the surface of the white dwarf, that is $\alpha_{\rm H} = 0$; (ii) when $\dot{M} = \dot{M}_{\rm up}$, $\alpha_{\rm H} = 1$.

The steady hydrogen burning converts hydrogen into helium on top of the C-O white dwarf. When its mass of the helium layer reaches a certain value, helium ignites. Helium shell burning is unstable if $\dot{M}_{\rm He} \le 10^{-6} M_\odot\ {\rm yr}^{-1}$, and a flash grows, during which part of the envelope mass is blown off (Kato, Saio & Hachisu 1989). For the helium accumulation ratio $\alpha_{\rm He}$, we use the results given by Kato et al. (1989).

Generally we have two kinds of products: (1) $M_{\rm WD}$ grows to 1.4 $M_\odot$, producing a SN Ia; (2) the calculations are stopped before $M_{\rm WD}$ reaches $1.4 M_\odot$ because $\dot{M}$ exceeds $\sim 10^{-4} M_\odot\ {\rm yr}^{-1}$ at which there is no wind solution, or $\dot{M}$ declines below $3\times 10^{-8} M_\odot\ {\rm yr}^{-1}$, at which a strong nova explosion is assumed to occur, or for a low-mass ($\le 1.5 M_\odot$) donor, the helium core mass reaches 0.46 $M_\odot$.

3. Results and discussion

An example of the calculated evolution ($M_{\rm WD,0} = 1.0 M_\odot$, $M_{\rm d,0} = 2.5 M_\odot$, $P_{\rm orb,0} = 1.6$ d) is plotted in Fig. 1. In this case the white dwarf can grow to $1.4 M_\odot$ to trigger a SN Ia. Despite of an initial mass ratio of 2.5, the

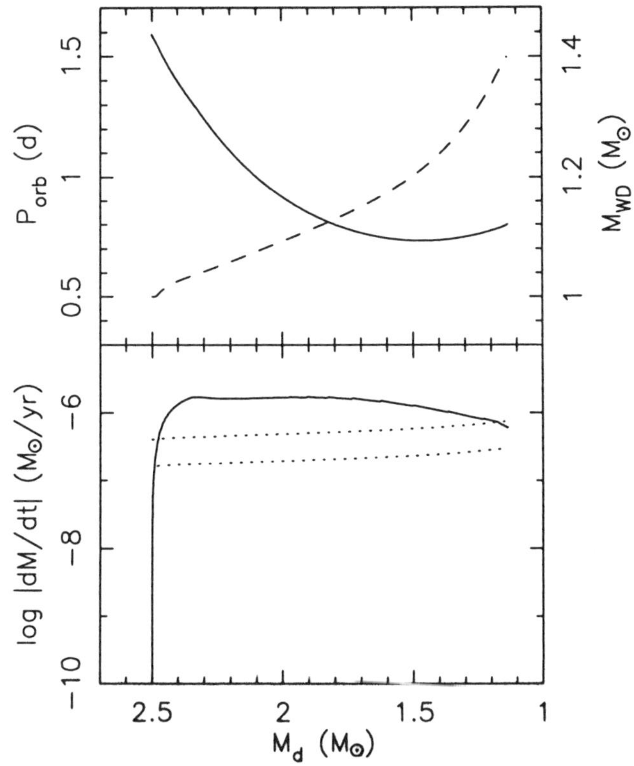

Figure 1. **Evolution of the white dwarf binary with $M_{WD,0} = 1M_\odot$, $M_{d,0} = 2.5M_\odot$ and $P_{orb,0} = 1.6$ d. Upper panel: solid and dashed curves denote the evolution of the orbital period and the mass of the white dwarf, respectively. Lower panel: evolution of the mass transfer rate. The dotted lines represent the steady burning region.**

mass transfer is stabilized by the strong wind, and maintains a rate around $10^{-6} M_\odot$ yr^{-1} during the whole evolution, i.e., nearly above the steady burning region. A large fraction ($\sim$ 70%) of the transferred matter is blown off in the wind. The orbital period decreases to ~ 0.7 d and then increases before the white dwarf mass reaches $1.4M_\odot$. A possible candidate that shows a similar evolutionary path is the transient supersoft X-ray source RX J0513.9-6951 in the Large Magellanic Cloud (Alcock et al. 1996; Cowley et al. 1996; Southwell et al. 1996).

Figure 2 presents another example of the evolution for $M_{WD,0} = 1.2M_\odot$, $M_{d,0} = 1M_\odot$, and $P_{orb,0} = 300$ d. The mass transfer rate, as in Fig. 1, is high enough to maintain stable hydrogen burning and weak helium flashes to increase the white dwarf mass. The orbital period increases as mass is transferred from the less massive donor to the white dwarf. Similar features can also be found in Fig. 2 in Hachisu et al. (1996).

We summarize the final outcome of our evolutionary calculations in

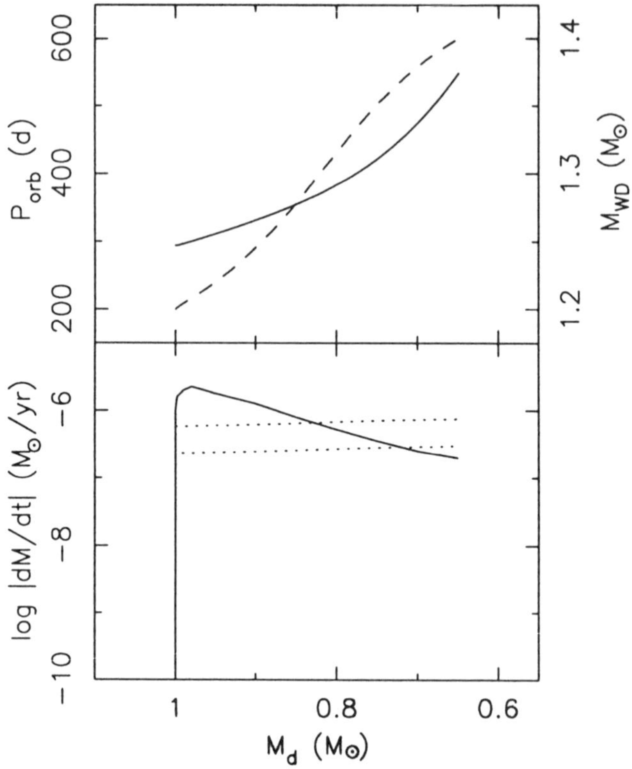

Figure 2. **Same as Fig. 1, but $M_{WD,0} = 1.2M_\odot$, $M_{d,0} = 1.0M_\odot$ and $P_{orb,0} = 300$ d.**

Fig. 3, showing the distribution of the progenitor systems of SNe Ia in the $M_{d,0} - P_{orb,0}$ diagram. It can be seen that there are two "islands", or two types of progenitor systems in the $M_{d,0} - P_{orb,0}$ diagram. One is close binaries that contain a massive ($M_{d,0} \sim 2 - 3.5M_\odot$) donor with an initial orbital period of several tenth of a day to several days, and mass transfer occurring in Case A and Case B (see dotted curves in the figure). The initial white dwarf mass that may increase to $1.4M_\odot$ in this case can actually be as low as $0.9M_\odot$. The other is low-mass ($M_{d,0} \sim 1M_\odot$) binaries with long orbital period (tens to hundreds of days), in which the donor is a red giant, i.e., in shell hydrogen burning phase, and the required initial mass of the white dwarf is $\sim 1.2M_\odot$, a bit larger than the value ($0.9M_\odot$) suggested by Hachisu et al. (1996).

Using the method suggested by Iben & Tutukov (1984), one can estimate the birth rate of SNe Ia, based on Fig. 3, to be $\sim 2\times10^{-3}$ yr^{-1} in our Galaxy, roughly consistent with the observed rate. The two types of progenitor systems in Fig. 3 can also qualitatively account for SN Ia explosions in both spiral and elliptical galaxies.

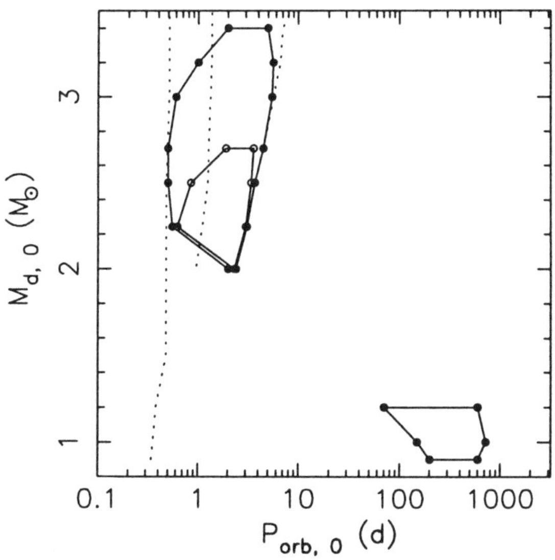

Figure 3. Distribution of the progenitors of SNe Ia in the $M_{\rm d,0} - P_{\rm orb,0}$ diagram. The filled dots denote the boundary of the initial orbital period at the beginning of the mass transfer, for a white dwarf of $1.2M_\odot$ initial mass with a specific companion star, and circles for $1M_\odot$ white dwarfs. The dotted lines represent the boundary of mass transfer in Case A and Case B (from left to right).

References

Alcock, C. et al. (1996) *MNRAS*, **280**, L49

Cowley, A. P. et al. (1996) in *IAU Symp. 165, Compact Stars in Binaries*, eds. J. van Paradijs, E. P. J. van den Heuvel & E. Kuulkers (Dordrecht: Kluwer), 439

Di Stefano, R. and Nelson, N. A. (1996) in *Supersoft X-ray Sources*, ed. J. Greiner (Berlin: Springer), 3

Eggleton, P. P. (1971) *MNRAS*, **151**, 351

Hachisu, I., Kato, M. and Nomoto, K. (1996) *ApJ*, **470**, L97

Hasinger, G. 1994, *Mod. Rev. Astron.*, **7**, 129

Iben, I. Jr. and Tutukov, A. V. 1984, *ApJS*, **54**, 335

Kahabka, P and Trümper, J. (1996) in *IAU Symp. 165, Compact Stars in Binaries*, eds. J. van Paradijs, E. P. J. van den Heuvel & E. Kuulkers (Dordrecht: Kluwer), 425

Kato, M., Saio, H. and Hachisu, I. (1989) *ApJ*, **340**, 509

Nomoto, K., Nariai, K. and Sugimoto, D. (1979) *PASJ*, **31**, 287

Prialnik, D. and Kovetz, A. 1995, *ApJ*, **445**, 789

Rappaport, S. A., Di Stefano, R. and Smith, M. (1994) *ApJ*, **426**, 692

Southwell, K. A. et al. (1996) *ApJ*, **470**, 1065

van den Heuvel, E. P. J. et al. (1992) *A&A*, **262**, 97
Yungelson, L. et al. (1996) *ApJ*, **466**, 890

ABUNDANCES FROM z=0 TO z=5

DONALD G. YORK
The University of Chicago
Department of Astronomy & Astrophysics
5640 S. Ellis Avenue
Chicago, IL 60637
USA

Abstract.

Interstellar abundances are compared for the Milky Way disk, the Milky Way halo, the Large and Small Magellanic Clouds and the damped Lyman alpha systems among the QSO absorption line systems. While a new set of observational aspects of element formation in the Universe is emerging, including a dearth of formation activity from z=5 to z=3, the predicted signal of [Si/Fe] decreasing from high z to low z, as Type I supernovae start contributing to Fe production, has not yet been seen.

1. Background

New telescopes and spectrographs now permit the study of the abundances of most of the first 30 elements in the periodic table in the interstellar gas. Following the determination of the first accurate abundances in gas clouds in the vicinity of the Sun, using the *Copernicus* telescope and spectrograph[(1)], modest resolution spectra of QSO absorption lines (QSOALS, the same resonance lines used in *Copernicus* studies, but redshifted into the 3500-9000 Åregion) were used to compare abundances for intergalactic gas clouds with those found in local Milky Way gas[(2)]. Observations with the *Hubble Space Telescope* allowed improvement and extensions of the former studies[(3)]. Observations with 4 meter telescopes[(4)(5),(6)] and, more recently, the Keck telescope[(7),(8)], have dramatically improved our understanding of abundances in the absorbers with $z_{abs} < z_{em}$ in spectra of QSOs.

K. Nomoto and J.W. Truran (eds.), Cosmic Chemical Evolution, 109–115.

I summarize here observations of local clouds, gas in the Small and Large Magellenic Clouds, and results from QSOALS from z=0.5 to 5. Particular emphasis is given to abundances of those species that, by virtue of column density, intrinsic line strengths, ionization properties and line placement, can be determined over a wide range of redshifts, of z, namely C, Si, Fe and Zn. For intercomparing gas at different z, it is convenient to measure abundances relative to zinc, referenced to solar relative abundances of the same elements. The relation [X/Y] refers to ratios by number of species X and Y (dex.) The notation $[X/Y]_o$ refers to the abundance referenced to the Sun.

2. The Galaxy

Steady improvements have occurred in our knowledge of abundances in local clouds. The improvements stem from better instruments and from improved knowledge of oscillator strengths (f-values.) In particular, for N^o, O^o, Si^+, Fe^+ and Ni^+ there exist lines weak enough to be unsaturated in local clouds with neutral hydrogen column densities N(H I) $< 10^{21}$ cm^{-2}. As has been known with low precision for some time, a pattern of depletion compared to solar abundances is seen in interstellar gas clouds which can be identified with three types of gas: cold clouds (ζ Oph B [(9)]), warm clouds (i.e. ζ Oph A [(9)], α Vir [(10)]), and halo clouds [(11)]. The most extreme depletions, interpreted as showing species of low gas density due to condensation on dust grains, occur in cold clouds[(12)] (Figure 1). The same pattern, but with successively less extreme depletion, ! ! occurs in warm clouds and in halo clouds. The depletions for the three types of gas and for selected elements are shown as solid lines in Figure 2[(13)].

Note that the depletions shown are with reference to Zn. This presentation is chosen because the abundances of high precision are based on column densities of first ions, the dominant stage of ionization for the species shown in an interstellar radiation field. Both H I and H II regions are represented in these measurements. Thus, a direct comparison with H is not possible as only N (H I) could be determined from absorption line observations. A literature search [(13)] reveals the errors in the ratios chosen here, owing to the different ionization potentials and to different local radiation fields, are less than ± 0.2. The log of the ratio N(Zn II)/N(H I) is shown at the far right. For regions represented here by the mean depletions shown, H I dominates over H II, but the relative amounts of H I and H II are unknown in detail.

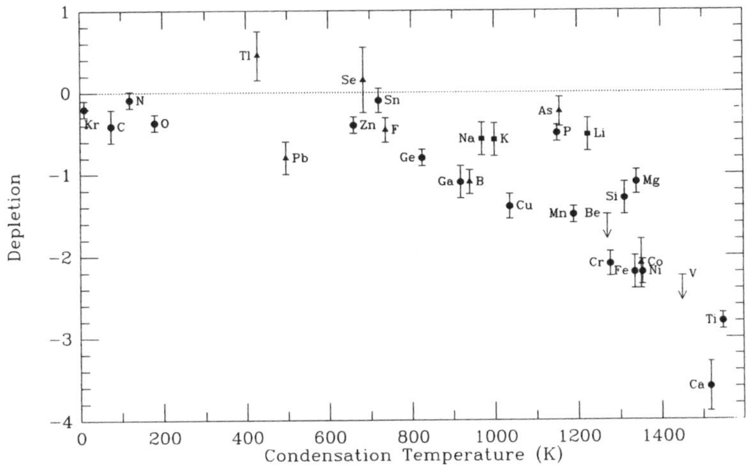

Figure 1. The variation of interstellar depletion for 30 elements with condensation temperature for an average dense cloud.

3. The Large and Small Magellanic Clouds

Depletions were determined for groups of lines (typically 4 - 5 components in 10 kms^{-1} wide complexes) toward Sk 108 in the SMC[(13)] and toward SN 1987A in the LMC[(14)]. Various means were used to show that, in all likelihood, H I regions are mainly represented by these depletions, though H II, in lesser quantities, exists alongside the H I, so much so that N II emission used in studying light echoes[(15)] can be correlated with the velocities of the absorption lines.

The results are shown by labelled symbols in Figure 2. The values are average depletions over five complexes toward Sk 108 and over eight complexes toward SN 1987A. No cold cloud pattern is formed toward SN 1987A, though narrow K I lines from gas with T$\leq$ 300 °K were observed with the AAT[(5)]. Either the column densities of dominant ions from the cold regions are much less than those from warm and halo clouds, or cold clouds in the LMC are depleted only as much as warm clouds in the Galaxy. This situation should be soon clarified by observation of the H_2 in the SMC and the LMC with the *FUSE* satellite. At any rate, the warm cloud and the halo cloud depletion pattern dominate LMC gas toward SN 1987A.

For Sk 108, only the halo cloud depletion pattern is seen: abundances, relative to Zn, of all elements studied are within 0.5 dex of solar.

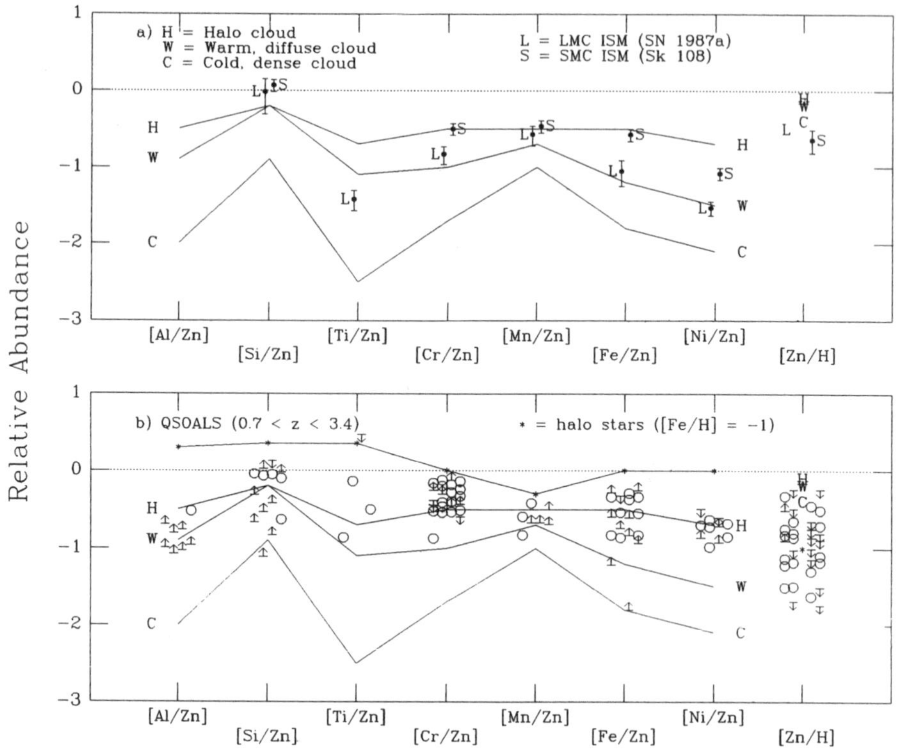

Figure 2. **Average abundances (solid or open circles) for systems of interstellar lines in the LMC/SMC (top) or QSO absorption lines (bottom). Solid lines connect mean abundances for three distinct types of interstellar clouds. Abundances are referenced to zinc (Zn II) to minimize ionization corrections for the first ions observed for the other elements shown. At the right is an indicator of total metallicity $[Zn/H]$ based on $[Zn\text{II}/H\text{I}] - [Zn/H]_{\odot}$.**

4. QSO Absorption Lines

4.1. DAMPED LYMAN ALPHA SYSTEMS

Some 40 damped Lyman alpha clouds have now been studied[(7),(8)]. Data summarized in references [(7)] and [(8)] are shown in Figure 2 averaged over all systems. Zn II is not generally detectable at the highest redshifts, explaining the upper limits at the highest redshifts. However, in halo type clouds (depletion wise), Si is essentially undepleted compared to Zn. Where both Si and Fe are observed, the trends are the same as for the averages in Figure 2.

Note again the values [Zn/H] at the far right. The absolute abundances of Zn consistently decrease with increasing z. The values of [Fe/H] in cold clouds in the Milky Way are similar to those in QSOALS, evidently because solar abundance local gas has depleted Fe into grains ([Fe/H]$\sim$0.001 [Fe/H]$_{\odot}$) whereas in many QSOALS, gas little depleted on grains has simply not reached solar abundance yet.

In the case of the QSOALS, it is not clear that the abundances are anything more than upper limits. Observations of Si II and O I in a few systems[(4),(16)] suggest that [O I/H I] referenced to [O/H]$_{\odot}$ is lower than [Si II/H I] referenced to [Si/H]$_{\odot}$. Assuming O and Si track each other owing to similar origins in type II SNe, the observations suggest that H II regions dominate over H I regions. (O I and H I track each other in ionization, so O I is largely destroyed in H II regions, whereas Si II remains a dominant ionization stage.) The alternative is that in the earliest SNe, Si is generated in greater masses relative to O in the ejecta than is the case at later times.

4.2. THE LYMAN ALPHA FOREST

Indications that there is a forest of C IV and Si IV doublets corresponding to the forest of Lyman alpha lines[(17)] and that [Si/C] is higher in these systems than in the local interstellar medium [(18)] have recently been confirmed [(19)] by extension to systems with N(H I) $\geq 10^{15}$ cm^{-2} and [N(H II)/N(H I)] $> 10^3$. The observations extend to $z = 4.5$, higher than the observations of damped Lyman alpha systems. Coupled with models of the ionization field from QSO radiation [(19),(20)], the data indicate a floor to heavy element abundances to the redshift limit so far observed [(21)].

5. Discussion

Gas phase abundance determinations indicate that heavy elements existed in QSOALS as early as $z = 5$, though at low levels. However, a marked increase in metallicity is noted beginning at $z = 3$. The results are consistent with an early universal contamination of primordial gas by massive stars, a delay in forming additional heavy elements until $z \sim 3$ [(22)] and a rise to 0.1 of solar abundances by $z = 2$. The abundances of Ca and Fe as seen in QSOALS do not show a further increase at $z < 1$ [(23)] to fully solar values, presumably because the disks analogous to the Milky Way in which solar abundance stars exist provide such a small cross section for absorption against background quasars, compared to the antecedent dwarf galaxies [(24)(25)] that may dominate at $z > 1$. Of course, as warm and cold cloud depletion patterns appear in low z galaxies, by analogy with the local interstellar medium, Ca, Fe and Si will be depleted, anoth! ! er possible explanation for the lack of solar abundances in low z QSOALS. As the poorly observed category of QSOALS from $z = 0$ to 1 is pursued [(6),(25)], more solar abundance systems should be seen in the species sulfur, zinc, oxygen and nitrogen (elements unaffected by depletion onto dust).

The data above should challenge and inform theoretical studies of nucleosynthesis as a function of time. For instance, the common observation of [Si/Zn]$\sim$[Si/Zn]$_o$ at high z, at times before SNe Ia could probably occur, suggests that Zn is a product of type II SNe [(26)], and not predominately associated with Type I SNe as some authors assume. The ratios [Si/C] above z =3 should aid in determining the mass of the first stars. The appearance of SNe at later times from successively less massive progenitors may be reflected in ratios [O/Si] in gas phase as a function of z with implications for the stellar mass function between $z = 1$ and $z = 3$. Further observations of Mn and Cu (odd-even nuclei) may show a measurable increase in ratios such as Mn/Fe and Cu/Fe as z increases [(27)]. Increased sample size may allow detection of the decrease in [α/Fe] (abundance of C, O, Si and other type II products compared to iron) with dec! ! reasing redshift. The excess is hinted at by comparison of [Fe/Si] in QSOALS (Fig. 2), but that result could be caused by residual dust depletion of Fe, in light of the fact that [Si/Zn] found in QSOALS is not seen to reach the value 0.4 found in old halo stars in the Milky Way [(28)].

I am indebted to my colleagues Priscilla Frisch, Richard Green, Lew Hobbs, Varsha Kulkarni, Jim Lauroesch, Dave Meyer, Jim Truran and Dan Welty whose shared verbal contributions equal the contributions noted in the references. David Achenbach and Chris Mallouris helped in the preparation of the paper.

References

1. Rogerson, J. B., Spitzer, L., Drake, J. F., Dressler, K., Jenkins, E. B., Morton, D. C., and York, D. G. 1973, *ApJ*, **181**, L97
2. Smith, H. E., Jura, M. and Margon, B. 1979, *ApJ*, 228, 369
3. Savage, B. D. and Sembach, K. R. 1996, *ARA&A*, **34**, 279
4. Kulkarni, V. P., Huang, K. L., Green, R. F., Bechtold, J., Welty, D. E., and York, D. G. 1996, *MNRAS*, **279**, 197
5. Pettini, M., Smith, L. J., Hunstead, R. W. and King, D. L.1994, *ApJ*, **426**, 79
6. Robertson, J., Morton, D. C., Blades, J. C., York, D. G. and Meyer, D. M. 1988, *ApJ*, **325**, 635
7. Lu, L., Sargent, W. L. W., Barlow, T. A., Churchill, C. W. and Vogt, S. S.1996, *ApJ*, **107**, 475
8. Prochaska, J. X.. and Wolfe, A. M. 1997, *ApJ*, **474**, 140
9. Sofia, U. J., Cardelli, J. A. and Savage, B. D. 1994, *ApJ*, **430**, 650
10. York, D. G. and Kinahan, B. F. 1979, *ApJ*, **228**, 127
11. Spitzer, L. and Fitzpatrick, E. L. 1995, *ApJ*, **445**, 196
12. Hobbs, L. M., Welty, D. E., Morton, D. C., Spitzer, L. and York, D. G. 1993, *ApJ*, **411**, 750
13. Welty, D. E., Lauroesch, J. T., Baldes, J. C., Hobbs, L. M. and York, D. G. 1997, *ApJ*, **489**, 672
14. Welty, D. E., Frisch, P. C., Sonneborn, G. and York, D. G. 1999, *ApJ*, in press
15. Xu, J., Crotts, A. P. S., and Kunkel, W. E. 1995, *ApJ*, **451**, 806
16. Lauroesch, J. T., Truran, J. W., Welty, D. E. and York, D. G. 1996, *PASP*, **108**, 641
17. Meyer, D. M. and York, D. G. 1987, *ApJ*, **315**, L5

18. Khare, P., York, D. G. and Green, R. 1989, *ApJ*, **347**, 627
19. Cowie, L. L, Songaila, A., Kim, T. and Hu, E. M.1995, *AJ*, **109**, 1522
20. Meiksin, A. and Madau, P. 1993, *ApJ*, **412**, 34
21. Songaila, A. 1997, *ApJ*, **490**, L1
22. Timmes, F. X., Lauroesch, J. T. and Truran, J. W.. 1995, *ApJ*, **451**, 468
23. Meyer, D. M. and York, D. G. 1992, *ApJ*, **399**, L121
24. York, D. G., Dopita, M., Green, R. and Bechtold, J. 1986, *ApJ*, **311**, 610
25. Ellis, R. 1999, *Nature*, xxx, xxx.(Astroph9807287).
26. Timmes, F. X., Truran, J. W., Lauroesch, J. T. and York, D. G. 1997, *ApJ*, **476**, 464
27. Truran, J. W. & Cameron, A. 1971, *ApJS*, **14**, 179
28. Wheeler, J. C., Sneden, C. and Truran, J. W. 1989, *ARA&A*, **27**, 279

ENRICHMENT OF HEAVY ELEMENTS IN CLUSTERS OF GALAXIES

YUHRI ISHIMARU
National Astronomical Observatory,
2-21-1, Mitaka, Osawa, Tokyo 181, Japan.
email: ishimaru@th.nao.ac.jp

Abstract. We study the roles of type Ia (SNIa) and type II supernovae (SNII) in enrichment of the intra-cluster medium (ICM). Considering stellar [Fe/H] of ellipticals, iron mass in ICM to luminosity ratios, and [α/Fe] of ICM, we show that $\sim 53\%$ of iron in ICM comes from SNIa. While clusters keep all of ejecta from ellipticals, we obtain the observational evidence for that groups of galaxies lose parts of the gas by supernovae-driven winds. This 'group wind' scenario is confirmed by an 1D hydrodynamical model.

1. Introduction

Clusters of galaxies are surrounded by hot ICM containing a large amount of heavy elements. Since the iron mass in ICM to total luminosity ratios (FeML) appears to take the universal value $\sim 0.01 - 0.02 M_{\odot}/L_{\odot}$, the iron in ICM should come from ellipticals (Arnaud et al. 1992; Renzini et al. 1993). We study the origin of iron in clusters, considering the shares of iron between ICM and ellipticals. Previously, since the stellar iron abundance of ellipticals was believed as high as $\sim 2Z_{\odot}$, the iron yield was regarded to be enough high to enrich the ICM mainly by SNII. Moreover, it was reported that ASCA observations gave high values of [α/Fe] to ICM (Mushotzky et al. 1996). These estimates seemed to support SNII enriched ICM. However, if all of iron in ICM is supplied only by SNII and SNIa products remain in halos of ellipticals, it contradicts to the observed iron abundance of halos which is lower than the solar value (e.g., Awaki et al. 1994). Thus, we reconsider these observational evidences, and then examine the roles of SNIa and SNII in iron enrichment of ICM, by using a chemical evolution

K. Nomoto and J.W. Truran (eds.), Cosmic Chemical Evolution, 117–121.

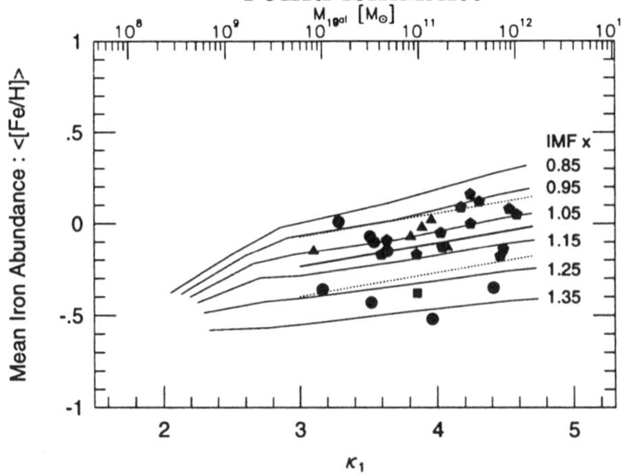

Figure 1. The mass-metallicity relation of ellipticals: The closed symbols indicate the observed relation based on the mean iron abundances. The tick solid and dotted lines indicate the regression line and the 1-σ dispersions, respectively. The thin solid lines give the theoretical relations for the IMF $x = 0.85$, 0.95, 1.05, 1.15, 1.25, and 1.35. The value of x is put at the next to the corresponding lines.

model based on a galactic wind scenario (cf. Arimoto & Yoshii 1987). In addition, combining X-ray data for ellipticals and groups of galaxies as well as clusters, we examine the evolution of groups and discuss differences in evolutional histories due to the mass scales of galaxy systems.

2. Clusters of Galaxies

2.1. MASS-METALLICITY RELATION OF ELLIPTICALS

The previous values of high stellar abundances of ellipticals were derived from the central Mg_2 index. However, since strong Mg_2 index gradient exists in an elliptical, these values must be over-estimated. Therefore, taking into account abundance gradients, we have estimated mean stellar abundances for 50 ellipticals (Arimoto et al. 1996). The obtained mean values are as low as $\sim Z_\odot$. Thus, using the mass-metallicity relation given by the new data, we have examined the iron productivity of ellipticals. As clearly shown in Fig.1, the observed relation requires the IMF slope x within a range of 1.10 ± 0.1, which is flatter than the Salpeter's IMF ($x = 1.35$).

2.2. IRON MASS IN ICM TO TOTAL LUMINOSITY RATIOS

The FeML indicates the efficiency of iron ejection from ellipticals. For the possible origins of iron in ICM, we consider i) SNII-driven galactic winds, ii) SNIa, and iii) stellar mass loss. We have calculated chemical evolutions of ellipticals with the various IMF slope; $x = 0.85 - 1.35$, and the various binary frequency, which determines SNIa rate; $b/b_\odot = 1/4 - 1$ (where, $b_\odot$

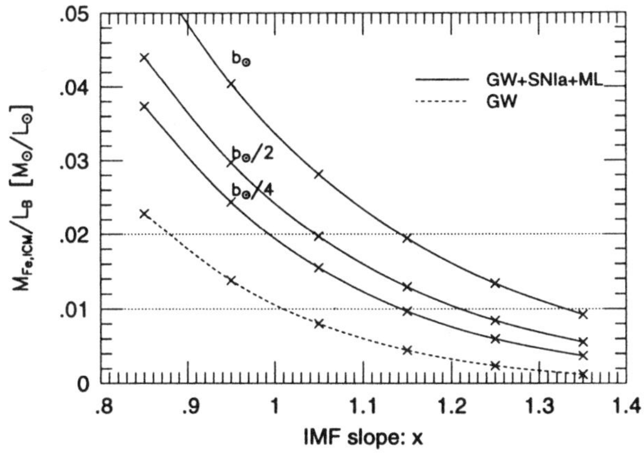

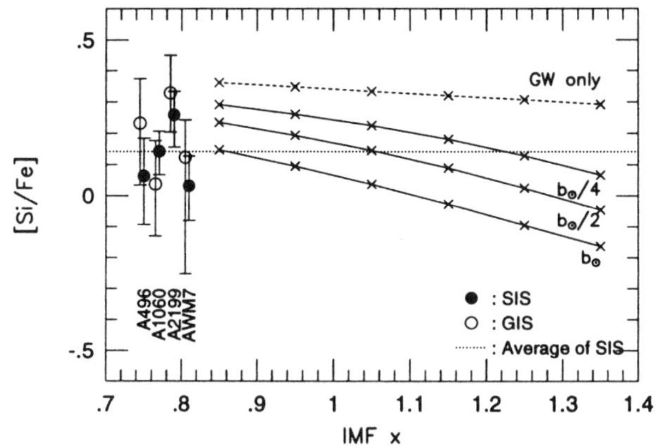

Figure 2. (upper) The relation between IMF slope x and FeMLs: The dashed lines gives the case that the ICM is enriched only by galactic winds. The solid lines indicate the case that stellar mass loss and SNIa also contribute to the ICM enrichment, as well as galactic winds. The values of assumed binary frequency are put at the next to the corresponding lines. The dotted lines gives the range of observational IMLR distributions.

Figure 3. (lower) The relation between IMF slope x and [α/Fe] of ICM: The solid and dashed lines indicate the same way with Fig. 2. At the left side, the ASCA data renormalized by meteoritic solar abundances are given. The dotted lines indicate the mean value of the ASCA data.

is that for the solar neighbourhood). The predicted FeMLs are given by integrations of iron ejection from each ellipticals. As shown in Fig. 2, the observed range of FeML cannot be explained only by galactic winds, unless IMF slope is as flat as $x \sim 0.9$, which causes too high stellar abundances. But if the ICM was enriched by SNIa as well as SNII, the FeML and the mass-metallicity relation can be explained consistently with the combinations of x and b; $(1.15 - 1.20, b_\odot)$, $(1.05 - 1.20, b_\odot/2)$, $(1.00 - 1.15, b_\odot/4)$.

2.3. RELATIVE ABUNDANCE RATIOS OF ICM

From recent ASCA observation, it has been reported that [α/Fe] of ICM seem to take over-solar values (Mushotzky et al. 1996). However, this discussion is based on comparison with the solar-photospheric abundances. A serious discrepancy is known to exist in iron abundance between the solar-photospheric value and the meteoritic value (Anders & Grevesse 1989). Thus, according to the prescription of supernovae nucleosynthesis by Nomoto et al., we have renormalized the ASCA data by the meteoritic values and examined the relative contributions of SNIa in ICM enrichment. As a result, the values of [α/Fe] decrease by ~ 0.16 dex and the fractional contribution of SNIa in iron is more than 50% (Ishimaru & Arimoto 1997). As shown in Fig.3, clearly the observed [α/Fe] requires SNIa products.

From these three observational constraints, we conclude that the $\sim 53\%$ of iron in ICM comes from SNIa. In this case, ellipticals have a flat IMF slope ($x \sim 1.1$) and a smaller binary frequency $b \sim b_\odot/2$.

3. Groups of Galaxies

3.1. ELLIPTICAL-GROUP-CLUSTER SEQUENCE

Combining X-ray data of clusters, groups, and ellipticals, we have obtained the new aspects in evolutions of smaller scale galaxy systems. Figure 4 shows the relation between gravitational mass $M_{\rm grav}$ and $M_{\rm gas}/L_B$. There is a critical mass at $M_{\rm grav} \sim 5\ 10^{13} M_\odot$, below which $M_{\rm gas}/L_B$ is strongly correlated with $M_{\rm grav}$. One can clearly identify the sequence from elliptical to clusters (hereafter EGC sequence). The EGC sequence is also established in the relation between $M_{\rm grav}$ and FeML with the same critical mass. If this correlation is due to accretions of baryons, FeMLs must take the universal value. The decreases of gas and iron suggests that smaller groups lose more intra-group medium (IGM). Especially, the EGC sequences are smoothly connected to ellipticals. Thus, we can infer that small groups eject most of IGM by supernovae-driven winds like ellipticals. The efficiency of group winds should be negatively correlated with the binding energy, and therefore, the EGC sequence would emerge.

3.2. GROUP WIND SCENARIO

We examine that group winds can indeed reproduce the EGC sequence under the assumption of the universal initial baryon fraction, using the 1D hydrodynamical model. The evolution of IGM is examined for groups with $L_B \sim 5\ 10^{10} - 5\ 10^{11} L_\odot$. The initial galaxy fraction is determined by the galaxy formation efficiency GFE= $M_{\rm galaxy}/M_{\rm baryon}$. The observed value of M/L_V; $\sim 70 - 150 M_\odot/L_\odot$ determines the possible range of GFE as $17\pm6\%$.

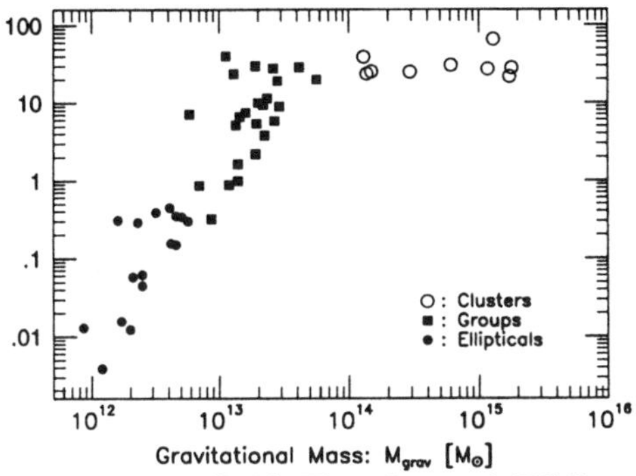

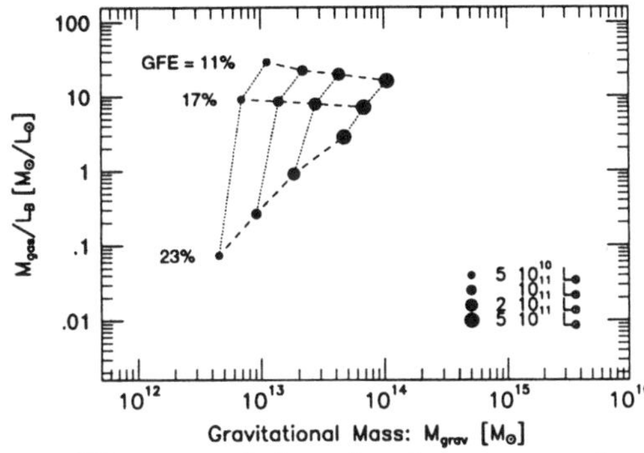

Figure 4. (left) The observed EGC sequence: The open circles, closed squares, and closed diamonds represent clusters, groups, and ellipticals, respectively.

Figure 5. (right) The calculated final gas-to-luminosity ratio as a function of gravitational mass: The total luminosities of each models are 5 10^{10}, 10^{11}, 2 10^{11}, 5 $10^{11} L_\odot$, which are given by smaller to larger closed circles. The dotted and dashed lines connect models with the identical luminosities and galaxy formation efficiencies, respectively. The values of GFE are put at the side of corresponding lines.

Figure 5 shows the rclation between the final M_{gas}/L_B ratio and total mass. As shown in this figure, a small dispersion in GFE causes a large dispersion in the efficiency of gas ejection. Because the GFE represent the energy source in a unit gas mass, groups with higher GFE eject more IGM. In addition, smaller groups is more sensitive to the GFE. Therefore, a possible dispersion in GFE easily generate the observed trend and dispersion.

If rich clusters are formed through hierarchical clusterings, the EGC sequence requires that *all* group wind ejecta come back successfully at merging of groups. On the other hand, if each galaxy system has evolved isolately, the EGC sequence can be understood simply by group winds. Moreover, unnatural baryon enhancement in rich clusters is not required.

References

Anders, E., & Grevesse, N. 1989, Geochim. Cosmochim. Acta, 53, 197

Arimoto, N., & Yoshii, Y. 1987, A&A, 173, 23

Arimoto, N., Matsushita, K., Ishimaru, Y., Ohashi, T., & Renzini, A. 1997, ApJ, 477, 128

Arnaud, M., Rothenflug, R., Boulade, O., Vigroux, L., & Vangioni-Flam, E. 1992, A&A, 254, 49

Awaki, H., Mushotzky, R. F., Tsuru, T., Fabian, A. C., Fukazawa, Y., Loewenstein, M., Makishima, K., Matsumoto, H., et al. 1994, PASJ, 46, L65

Ishimaru, Y., & Arimoto, N. 1997, PASJ, 49, 1

Mushotzky, R., Loewenstein, M., Arnaud, K. A., Tamura, T., Fukazawa, Y., Matsushita, K., Kikuchi, K., & Hatsukade, I. 1996, ApJ, 466, 686

Renzini, A., Ciotti, L., D'Ercole, A., & Pellegrini, S. 1993, ApJ, 419, 52

ON THE ORIGIN OF METALLICITY IN LYMAN-ALPHA FOREST SYSTEMS

M. CHIBA
National Astronomical Observatory, Mitaka, Tokyo 181, JAPAN
AND
B. B. NATH
Raman Research Institute, Bangalore-560080, INDIA

1. Introduction

Several recent observations indicate that even the low H I column density Lyα absorption lines may not be primordial, as they were previously thought to be. High-resolution observations using the Keck telescope (Cowie et al. 1995; Tytler et al 1995.; Songaila & Cowie 1996) have revealed that a large fraction of Lyα absorption lines with H I column density as low as $N_{HI} \sim 10^{14}$ cm^{-2} at redshift $z \sim 3$ are associated with metal lines (C IV). This fraction, hereafter denoted as f_m, is estimated as 0.5-0.75. On the other hand, there is an indication that at low z a large fraction of Lyα absorption lines are directly associated with halos of visible galaxies. This fraction is hereafter denoted as f_g. Lanzetta et al. (1995, hereafter LBTW) claimed $f_g \sim 0.6$ at $z \lesssim 1$ (see also Bowen, Pettini & Boyle 1997).

In the hierarchical model of structure formation, there is a continuum of halos of various masses and velocity dispersions. In this contribution, we consider the hypothesis that the Lyα absorption lines are produced in two populations of halos in this continuum – in 'minihalos' of small velocity dispersions that have never cooled and formed stars, and in large galactic halos with stars and metals. Based on this two-population model for Lyα absorbers, together with chemical history of galactic halos, we investigate whether or not f_m is understood in terms of f_g by the relationship $f_m \simeq f_g$. If f_m is larger than f_g, it may be concluded that the intergalactic medium (IGM) experienced a global enrichment at higher z.

K. Nomoto and J.W. Truran (eds.), Cosmic Chemical Evolution, 123–127.

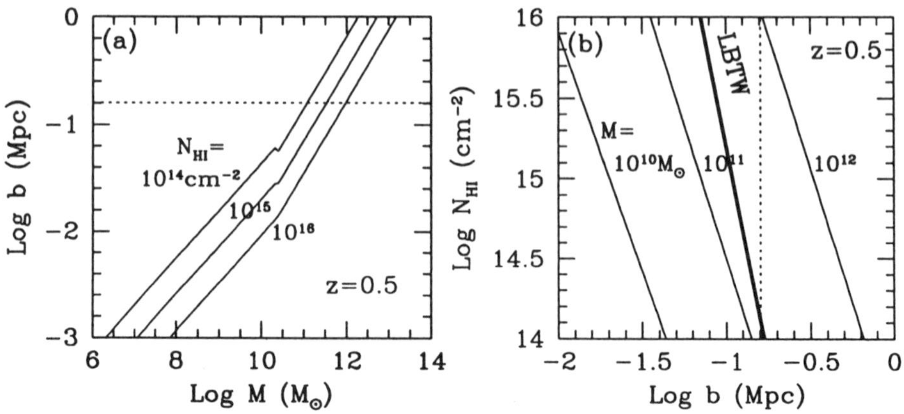

Figure 1. (a) The impact parameter b (Mpc) of a sightline as a function of the total mass of a halo M ($M_\odot$) to produce an H I column density N_{HI} at $z = 0.5$. (b) The H I column density N_{HI} (cm^{-2}) against the impact parameter b (Mpc).

2. Two-population model for Lyα absorbers

We assume that the density profile of both minihalos and galactic halos is represented by that of a softened isothermal sphere, which is characterized by the circular velocity V_c. We define minihalos as halos in which baryonic gas is entirely photoionized by the external UV background and the gas is stably confined in the gravitational field of the dark matter (Rees 1986). Minihalos are bounded by two scales of V_c, $V_1 < V_c < V_2$, where the lower limit V_1 is set by the sound velocity of the photoionized IGM, which is $\sim 15\,(T_{IGM}/10^4\,\mathrm{K})^{0.5}$ km s^{-1}. The upper limit V_2 is taken as ~ 55 km s^{-1}, following the argument by Thoul and Weinberg (1996) on galaxy formation under the UV background. We will call the halos with $V_2 < V_c < V_3$ galactic halos. Our results below are not sensitive to the upper limit V_3, and we will use a value of 250 km s^{-1}. In this case, Lyα absorbers in galactic halos are assumed to be confined with ambient hot halo gas (Mo 1994).

Figure 1(a) shows the relation between the impact parameter b and the total mass of the system M when N_{HI} is given. It is seen that the H I column densities as low as 10^{14} cm^{-2} can be provided by either the low-mass minihalos with $b \lesssim 40$ kpc or by the high-mass galactic halos with larger b. Figure 1(b) shows N_{HI} vs. b for a given total mass. The dotted line denotes the observed size of gaseous envelopes $\sim 160h^{-1}$ kpc in luminous galaxies that give rise to Lyα absorption lines at $z \lesssim 1$ (LBTW). The thick line indicates the observed correlation between N_{HI} and b obtained by LBTW. These results suggest that the current model reproduces the observational results in a reasonable manner.

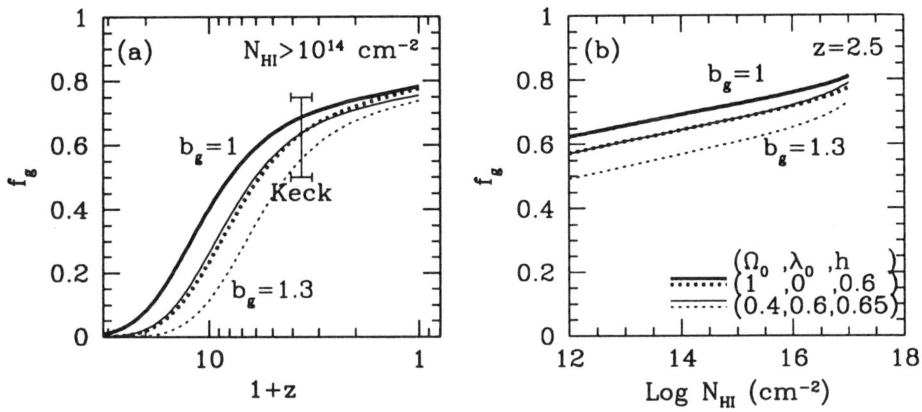

Figure 2. (a) The redshift evolution of the fraction of Lyα absorption lines associated with galactic halos, f_g, in the range of $55 \leq V_c \leq 250$ km s^{-1} (while $15 \leq V_c \leq 55$ km s^{-1} for minihalos), with $N_{HI} > 10^{14}$ cm^{-2}. The error bar denotes the Keck results for the fraction of Lyα lines associated with CIV metal lines. (b) The dependence of f_g on N_{HI} at $z = 2.5$.

3. The number of Lyα absorption lines

Based on the Press-Schechter formalism for deriving the mass function of collapsed halos, we calculate the number of Lyα absorption lines per unit redshift dN/dz in the universe, which is specified by cosmological parameters $(H_0, \Omega_0, \lambda_0)$. The power spectrum of density perturbations is taken from the CDM, with a bias parameter b_g for normalization. It is found that the value of dN/dz decreases with decreasing z as the hierarchical mergings proceed. Detailed comparison with observations suggests that the model reproduces the observed number of Lyα lines, as long as V_1 is about 15 km s^{-1} (see Chiba & Nath 1997 for details).

We then evaluate the fraction of Lyα absorbers associated with galactic halos, f_g, by calculating the number of absorbers in the range $V_2 < V_c < V_3$ for galactic halos. Figure 2(a) shows the redshift evolution of f_g for lines with $N_{HI} > 10^{14}$ cm^{-2}, where we assume $V_2 = 55$ km s^{-1}. It is found that at high z ($z \gtrsim 10$), most of the Lyα lines are provided by minihalos that collapsed early in the CDM power spectrum. The fraction f_g then increases with time as a result of the formation of massive halos via hierarchical merging. It is worth noting that the predicted value of f_g at $z \sim 2.5$ are reasonably in agreement with the reported value of f_m for metal lines. We also obtain the gradual increase of f_g with increasing N_{HI} [Figure 2(b)], which is to be confirmed from observations.

4. Metallicity of Lyα absorbers

Several recent observations have also measured the metal abundances of Lyα lines at high z (Cowie et al. 1995; Womble et al. 1996). The presence of heavy elements implies that interstellar gas confined in dark halos has been processed by star formation and subsequent chemical evolution. Here we take the following simplified approach to evaluating the metallicity of Lyα lines associated with galactic halos. First, absorbers confined with gaseous halos are assumed to be enriched as a result of formation of metal-poor halo stars and/or globular clusters. According to observations and models of these population II objects in the Galaxy (Schuster & Nissen 1989; Pagel 1989), the rate of metal enrichment is of the order of 10^{-3} Gyr^{-1} from the possible age-metallicity relationship, and the process of enrichment lasted for a few Gyrs from the observed spread in ages. Second, star formation from gas components is assumed to start when background dark halos collapse. The latter epochs are evaluated from typical epochs of collapse for 1σ density fluctuations.

Figure 3 shows the evolution of the mean abundance for absorption lines with $N_{HI} > 10^{14}$ cm^{-2} (solid and dotted lines). It is found that the model reasonably reproduces the Keck results of [C/H]$= -2.5 \sim -2$ at $z \sim 2.5$. We assume $\log < Z > /Z_{\odot} = -1$ as an upper limit for abundance of absorption lines, since the subsequent formation of halo stars may be inhibited as a result of gas accretion onto the disk component. Also shown in the figure with filled circles are the recent findings of Lu et al. (1996) for the abundances ([Fe/H]) of damped Lyα systems observed with Keck. Intriguingly, the median values of abundances in such systems are well reproduced by the current models. This may support the hypothesis, first proposed by Lanzetta, Wolfe & Turnshek (1995), that damped Lyα systems could represent a spheroidal component of galaxies.

5. Concluding remarks

We have shown that a fraction of Lyα forest lines with associated metal lines can be understood in the framework of hierarchical structure formation model, assuming that halos with circular velocity $15 \lesssim V_c \lesssim 55$ km s^{-1} do not produce stars and heavy elements. The model predicts that the fraction of Lyα lines with associated metal lines increases slowly with the H I column density. We have further used the age-metallicity relation of Galactic halo stars to predict the evolution of the average metallicity of the associated metal line systems. In particular, we find that the average metallicity at $z \sim 2$ should be about 1/100th of solar.

As an observational test for this model, we propose to determine the firm limits on the fraction f_m of Lyα lines associated with metal lines,

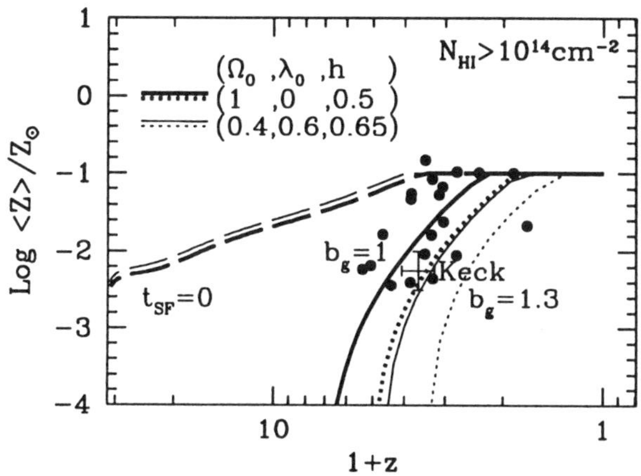

Figure 3. The redshift evolution of the average metallicity $< Z >$ of absorption lines based on the model described in the text. For reference, the metallicity evolution if the star formation commences from the beginning of the universe, $t_{SF} = 0$, are shown as dashed lines.

dependence of f_m on z and N_{HI}, and dependence of metallicity in Lyα absorbers on z.

References

Bowen, D.V., Pettini, M., & Boyle, B.J. 1997, MNRAS, in press (astro-ph/9707306)
Chiba, M. & Nath, B.B. 1997, ApJ, 483, 638
Cowie, L.L., Songaila, A., Kim, T.S., & Hu, E.M. 1995, AJ, 109, 1522
Lanzetta, K.M., Wolfe, A.M., & Turnshek, D.A. 1995, ApJ, 440, 435
Lanzetta, K.M., Bowen, D.B., Tytler, D., & Webb, J.K. 1995, ApJ, 442, 538 (LBTW)
Lu, L., Sargent, W.L.W., Barlow, T.A., Churchill, C.W., & Vogt, S.S. 1996, ApJS, 107, 475
Mo, H.J. 1994, MNRAS, 269, L49
Pagel, B.E.J. 1989, Rev. Mexicana Astron. Astrofis., 18, 161
Rees, M.J. 1986, MNRAS, 218, 25P
Schuster, W.J. & Nissen, P.E. 1989, A&A, 222, 69
Songaila, A., & Cowie, L.L. 1996, AJ, 112, 335
Thoul, A.A. & Weinberg, D.H. 1996, ApJ, 465, 608
Tytler, D., Fan, X.-M., Burles, S., Cottrell, L., Davis, C., Kirkman, D., & Zuo, L. 1995,, in QSO Absorption Lines, ed. G. Meylan (Berlin: Springer-Verlag), p. 289
Womble, D.S, Sargent, W.L.W., & Lyons, R.S. 1996, in Cold gas at High Redshift, eds. M. Bremer, H. Rottgering, P. van der Werf, C. Carilli (Kluwer), in press

A METAL ENRICHED DARK CLUSTER OF GALAXIES AT $Z = 1$

Dark lens searches

M.HATTORI
Astronomical Institute, Touhoku University
Aoba Aramaki, Aoba, Sendai 980-77, Japan

1. Introduction

Looking for and studying very distant galaxy clusters, clusters at $z > 1$, are one of the prime subjects of the modern observational cosmology. If the metallicity of the hot intra-cluster medium in very distant galaxy clusters is measured for example, it provides fruitful informations for us to understand the formation and evolution of galaxies. However, difficulty of the study is that there is few confirmed very distant galaxy clusters yet. We first have to search for very distant clusters but it requires very deep observations. A random selection of sky is not practical. We have to select the sky. In this article, it is demonstrated that missing lens problem has close connection with very distant cluster of galaxies and dark lens searches could open a new window for studying very distant cluster of galaxies.

2. Missing lens problem

The word 'missing lens problem' can be found in the unique text book on the gravitational lensing written by Schneider, Ehlers, Falco (1992). It is the problem that there are multiple QSOs which are suspected to be gravitational lens systems, but for which no deflector is observed near the images or only a part of deflector is observed. The difference of the latter cases from the usual missing mass problem is that it requires unusually high mass-to-light ratio compared with that of already known objects. An yet unidentified part of the lens objects as already known objects, such as galaxy and/or galaxy clusters, is called 'dark lens'. The dark lens search is the trial to identify the dark lens objects as already known objects (e.g. Wiklind & Combes 1996). If all the trials in multi wave length with enough deepness failed to find any evidence of already known objects as the lens,

K. Nomoto and J.W. Truran (eds.), Cosmic Chemical Evolution, 129–138.

it could lead us to conclude the existence of an yet unknown new type of object which contains few luminous matter.

A Table 2 summarizes the candidates of the multiply and ring imaged quasar and radio lenses. We can see that a rather large fraction of the lens candidates are showing the missing lens problem.

3. X-ray search for dark lens objects

A dark lens object could be a bright X-ray source since it may accumulate the inter-galactic medium by its strong gravitational field and heat the accumulated gas up to a temperature of a few keV due to the gravitational energy release, especially for the cases of the image separation of a few arcsec or more. Therefore, a deep pointed X-ray observations could be a promising method for identifying the lens objects complemental to optical and infrared searches. Even if the deep X-ray observations failed to find the lens objects, it provides strict upper limit on the fraction of the hot X-ray emitting gas mass in the dark lens objects which is one of the dominant components of luminous matter in the nearby known objects. Non detection in X-ray, therefore, let us cast toward an exciting possibility, that is existence of a dark matter condensation containing few luminous matter.

Motivated by those, we have been performing X-ray search for dark lens objects by using the X-ray satellite ASCA and ROSAT. In the following sections, our results obtained for two objects are summarized.

4. MG2016+112:A dark cluster

MG2016+112 was first discovered as triple radio sources by Lawrence et al. (1984). A part of lenses have been identified as two galaxies, galaxy D at $z = 1.01$ (Schneider et al. 1985 & 1986) and galaxy C (Lawrence et al. 1984). However, only the luminous matter of these galaxies is not enough to explain the observed nature of the lensed images. Narasimha, Subramanian & Chitre (1987) proposed a gravitational lensing model that the quasar MG2016+112 is being lensed by the two galaxies assisted by a cluster of galaxies centered on the galaxy D. However, it was reported that no evidence of the existence of the postulated cluster have been found in spite of deep optical (Schneider et al. 1985) and infrared (Langston et al. 1991) search. We have performed deep X-ray observations of MG2016+112 and found evidences of a cluster of galaxies at $z = 1$ (Hattori et al. 1997).

4.1. THE ASCA OBSERVATIONS

A deep observation of MG2016+112 with the ASCA X-ray observatory was performed from October 24th to 26th 1994, with an effective exposure time of 85 ksec. A new X-ray source (referred to as AXJ2019+1127) was discovered in the direction of the lens system with more than 10σ significance level. Figure 1 shows the X-ray spectrum of AXJ2019+1127 obtained by ASCA. An emission line feature is clearly seen at 3.5 keV which matches with the redshifted Fe K line from a cluster of galaxies at $z = 1$. Assuming that the source is a cluster, the spectrum can be fit with a Raymond-Smith plasma model (Raymond & Smith 1977). The best fitting results are summarized in Table 1. The inclusion of a non-zero abundance reduces χ^2 by 10.01 and is significant at the 99.9% level (3σ). The best fit value of N_H is consistent with the Galactic value of $N_H = 0.15 \times 10^{22}\mathrm{cm}^{-2}$ (Dickey & Lockman 1990). As shown in Fig.3, the observed X-ray luminosity and temperature of AXJ2019+1127 is consistent with the correlation of X-ray gas temperature and X-ray luminosity for distant clusters (Mushotzky & Scharf 1997). These results are consistent with the idea that there is a cluster of galaxies including the galaxy D as a member galaxy.

TABLE 1. The fitting results for the ASCA spectrum of AXJ2019+1127 by Raymond & Smith model. Errors are the 1σ confidence level statistical errors for one interesting parameter except the errors for the redshift where the $\pm 1\%$ systematic error in the ASCA energy scale calibration is included in the 90% statistical errors. The solar iron abundance of 4.67×10^{-5} is adopted (Anders & Grevesse 1989) and throughout the paper except in Fig.3 $H_0 = 50h_{50}$km/s/Mpc and $q_0 = 0.5$ are assumed.

kT keV	Z $Z_\odot$	$L_x(2-10\mathrm{keV})$ $10^{44}h_{50}^{-2}$erg/sec	z	N_H cm^{-2}	$\chi^2/d.o.f.$
$8.6^{+4.2}_{-3.0}$	$1.7^{+1.25}_{-0.74}$	$8.4^{+2.4}_{-1.7}$	$0.94^{+0.08}_{-0.09}$	$0.18^{+0.15}_{-0.10} \times 10^{22}$	$46.66/52 = 0.90$

4.2. THE ROSAT HRI OBSERVATIONS

Another crucial test for the cluster origin of the X-ray emission would be to resolve the source extent. This was performed with a deep observation with the ROSAT HRI. We find a diffuse source with 76 ± 24 source counts, $(1.4 \pm 0.4) \times 10^{-3}$ cts/sec within a circle of radius $1'$. Fig.2 shows the source image. The peak detection significance of the source is 4.2σ. The maximum of the X-ray emission is consistent with the position of the cD galaxy D within the HRI central position determination accuracy, at most $\sim 20''$, limited both by pointing accuracy and poor photon statistics. The HRI count rate is consistent with, but slightly less than that expected from the ASCA observation, that is $\sim 2.5 \times 10^{-3}$cts/sec. There is no other source above the 3σ significance detection limit in the circle of radius $3'$ centered on MG2016+112 in the HRI field. We therefore conclude that the

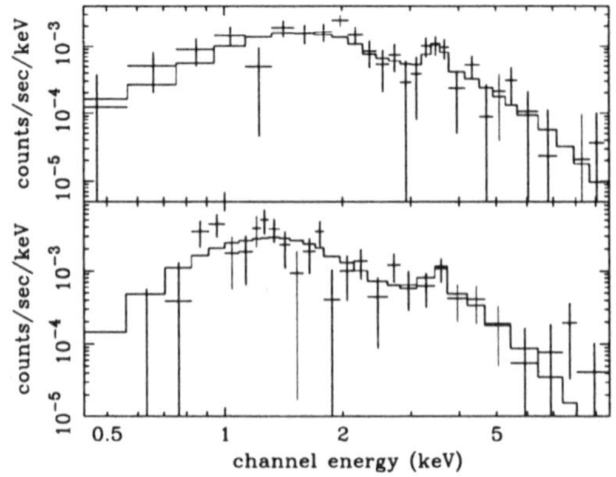

Figure 1. X-ray spectrum of AXJ2019+1127 obtained by ASCA. The background subtracted source spectrum obtained by two GISs (top panel) and by two SISs (bottom panel) with 1σ Poisson errors are shown. The solid lines are the best fit Raymond-Smith models obtained by simultaneous fitting of the GIS and SIS spectra. The background is taken from blank field observations during the test (PV) phase of the ASCA mission. The SIS data were taken with 2CCD bright mode observations with effective exposure time of 22ksec.

source detected by the ROSAT HRI is identical to AXJ2019+1127. Fig.2 shows that the X-ray surface brightness distribution of the source deviates significantly from the profile of the ROSAT HRI point spread function and the source is extended. Since the source counts within a concentric circle monotonically increases up to $1'$, the source radius is estimated to be $1'$ ($500h_{50}^{-1}$kpc for $z = 1$).

4.3. THE MOST DISTANT X-RAY CLUSTER AS A DARK CLUSTER

The above results clearly imply that AXJ2019+1127 is a galaxy cluster in which galaxy D is the central dominant galaxy. This is the most distant galaxy cluster discovered in X-rays so far. Since the X-ray emission provides the best evidence that clusters are gravitationally bound entities (unless a large number of measured galaxy redshifts are available), this is also the most distant clearly confirmed galaxy cluster. The redshift of the next furthest known X-ray cluster is MS1054.5-0321 at $z = 0.826$ (Donahue et al. 1997). Assuming that the cluster is isothermal and in hydrostatic equilibrium, one can obtain a rough estimate of the cluster mass. The best fit results yields a central electron density of $1.7 \times 10^{-2} h_{50}^{0.5}$ cm^{-3}, a gas mass of $2.4 \times 10^{13} h_{50}^{-2.5} M_\odot$ and a gravitational mass of $3.6 \times 10^{14} (kT/8.6\mathrm{keV}) h_{50}^{-1} M_\odot$ within a sphere of radius of $500h_{50}^{-1}$kpc in the cluster. The gas mass fraction of $\sim 7h_{50}^{-1.5}$% of the total mass is typical for the central regions of nearby clusters (Briel et al. 1992) within the measurement errors.

The obtained mass distribution is consistent with the mass model for

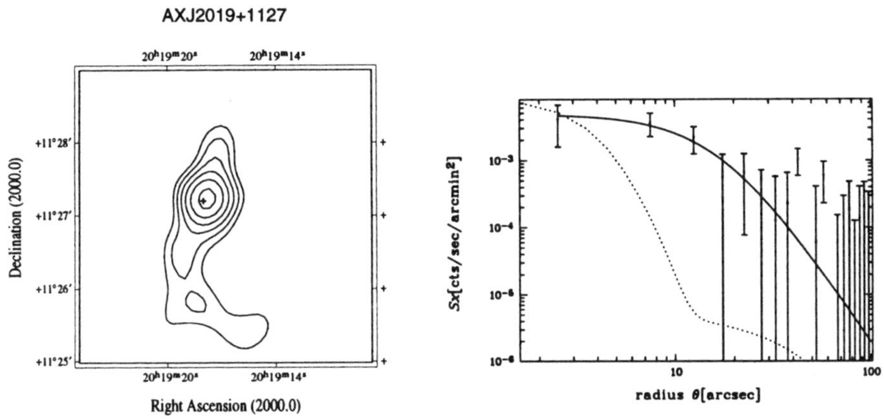

Figure 2. **a.** The left pannel shows that X-ray image of AXJ2019+1127 obtained by the ROSAT HRI. A deep ROSAT HRI observation was performed to resolve the possible extent of the source. The lens system was observed from November 15th to 20th 1995, from April 19th to 28th 1996 and from October 22nd to November 13th 1996 with a total exposure time of 79 ksec, yielding an effective exposure of 53.7 ksec after removal of observing times with high background flux. The accuracy of the HRI pointing is calibrated with two point sources coincident with stars in the Palomar Sky Survey observed in the field. The accuracy of the HRI pointing is better than $9''$. The contours show the significance levels with a linear spacing of 0.5σ from the minimum level of 1.5σ. The peak significance is 4.2σ. The image is smoothed with Gaussian filter with $\sigma = 15''$. The cross shows the position of galaxy D. **b.** The right pannel shows that the background subtracted X-ray surface brightness profile of AXJ2019+1127 obtained by the ROSAT HRI. The ROSAT HRI PSF profile at the same off-axis angle as the source position is also shown (dashed line). The central surface brightness of the PSF is set for a point source at the same central surface brightness as the source. Superposed on the data is the best fit isothermal β model (solid line) with a core radius of $17'' \sim 150h_{50}^{-1}$kpc, and $\beta = 0.9$ (solid line) which is typical for X-ray profiles of nearby rich clusters (Sarazin 1986). The central surface brightness of the model is $\sim 3.3 \times 10^{-3}$cts/sec/arcmin^2.

the lens proposed by NSC (model II in their paper). It implies that we have indeed discovered cluster responsible for the lensing effect additional to galaxy D. Another spectacular result of this discovery is the high iron abundance detected in the intra-cluster medium of AXJ2019+1127. Although the errors on the abundance are quite large, we can safely say that the iron content in this cluster is at least as high as that for nearby clusters of galaxies (Ohashi 1995). Currently the most distant galaxy cluster in which an Fe K-line has been detected is at a redshift of $z = 0.54$ (Donahue 1996). Therefore, AXJ2019+1127 is now the highest redshift cluster for which the existence of a metal enriched intra-cluster medium is confirmed. The detection of the large iron content at high redshift sets a new limit for the epoch of the enrichment. This favors models where early star burst phases in galaxies are responsible for the metal enrichment of the intra-cluster medium (Arnaud et al. 1992; Hattori & Terasawa 1993).

More surprisingly, a high iron abundance has been discovered in this "dark cluster", which is very poor in its galaxy content according to the deep optical and infrared searches (Schneider et al. 1985; Langston et al. 1991). The main optical light source from this cluster is galaxy D which has a blue luminosity of $L_B = 1.1 \times 10^{11} h_{50}^{-2} \mathrm{L}_{\odot}$. Another possible member galaxy is galaxy C (Lawrence et al. 1984) which, however, can contribute only a fraction of galaxy D in optical luminosity. Therefore, the mass to light ratio of the cluster would be $M/L_B \sim 3300(kT/8.6\mathrm{keV})^{-1} h_{50} M_{\odot}/L_{\odot}$! Since according to our current understanding the metal enrichment originates in the stars of the cluster galaxies, it is very puzzling that we detect such a high iron abundance in this "dark cluster". Therefore, either AXJ2019+1127 is a very new and enigmatic object or it has more cluster members at fainter magnitudes that have escaped the deep searches conducted so far. Recently, it has been reported (Ebeling et al. 1995) that there are also some low redshift clusters that are optically dark but bright in X-rays. Combined with our results, this may indicate that there exist many such objects in our universe which have eluded optical identification. In any case, AXJ2019+1127 provides us with a new means for testing cosmological theory and ought to be well studied in various wavelength bands.

4.4. THE MASS DISTRIBUTION OF THE CLUSTERS AT $Z \sim 1$

The radio observation has confirmed that there is no additional images of the lensed quasar MG2016+112 more than $10''$ and less than $2'$ from the brightest image down to two order of magnitude fainter than the brightest image (Schneider et al. 1985). This gives a strong constraint on the mass distribution of AXJ2019+1127 so that the Einstein ring radius of the cluster against the sources at $z = 3.27$ is less than $10''$ or $20''$ if the possibility that the galaxy D is off-center of the cluster within the central determination accuracy, is taken into account. Further observations of AXJ2019+1127 to determine the center of the cluster with very good accuracy may provide an unique opportunity to study the mass distribution of this very high redshift cluster and test the structure formation theory (Navarro, Frenk, White 1995).

5. Q2345+007:A cold cluster

The quasar Q2345+007 has been one of the most mysterious lens candidate double quasar since its discovery (Weedman, et al. 1982). In spite of deep and wide field optical searches for lens object, main lens object has not yet been identified (e.g. Tyson et al. 1986). A very faint galaxy was found edge of the secondary image, B image, after the subtraction of the point spread function of the image (Fischer et al. 1994). Since a C IV doublet absorption line ($z = 1.483$) (Foltz et al. 1984, Steidel & Sargent 1991) was found in the

B image, the redshift of the faint galaxy is supposed to be 1.483. However, the expected mass of the galaxy is too small to explain the wide separation of the two images unless it has an extremely high mass-to-light ratio. A large cluster at $z = 1.49$ was claimed since both of two images have metal absorption lines with redshift $z = 1.491$. Bonnet et al. (1993) reported the detection of a possible lensing cluster from weak lensing shear field with a strength of $\gamma \sim 0.15$, confirmed by van Waerbeke et al. (1997) and arclet candidates. After this detection, a galaxy cluster candidate was found as an enhancement in the number density of faint galaxies at the right location on the sky predicted by the shear field (Fischer et al. 1994, Mellier et al. 1994). Pello et al. (1996) made photometric redshift estimations for the galaxies in the cluster candidate field and found an excess of galaxies at $z \sim 0.75 \pm 0.1$. Assuming the cluster has this redshift, the velocity dispersion required to produce the observed shear pattern should be 790^{+115}_{-170}km/s (Pello et al. 1996) which corresponds to the hot gas temperature of $kT = 4.1^{+1.3}_{-1.6}$keV. However, the cluster with this velocity dispersion centered at $40''$ off from the brightest image can not be the main lens for Q2345+007.

The 46.3ksec exposure of the ROSAT HRI observation and the 80ksec ASCA observations did not detect an X-ray emission from the cluster responsible to the observed shear. It sets 3σ upper limit on the X-ray luminosity of 0.4×10^{44}erg/sec in rest frame $2 - 10$keV (Hattori et al. 1997c). In Fig.3, this upper limit and the temperature range expected from the shear field measurement are plotted. Upper limit of the X-ray luminosity is rather small compared with the value expected from its temperature. The cluster may contain little hot gas. However, the errors in the temperature estimation is large because of the simple modeling of the cluster mass distribution, unknown source redshifts and large errors in shear strength measurement and the cluster redshift estimation. Therefore, further observations to constrain the cluster mass from lensing and the spectroscopical measurement of the cluster redshift are very important.

Up to now, dark lens searches for Q2345+007 in multi-wave length have failed in identifying main lens objects as galaxy and/or galaxy cluster. It supports the idea that this double quasar is physically associated and is not lensed single quasar (Kochanek et al. 1997). Kochanek et al. (1997) is arguing that the most of the wide separation angle double quasars without lens could be physical association. However, we have to remind ourself the remarkable similarity of spectra between two images of the double quasars which is of course naturally explained by lensing, compared with the non-similarity of randomly selected two quasars' spectra (Simcoe, Richards, & Turner 1997). If the most of the double quasars are turned out to be physical pair, it requires a new model which controls physical state of two quasars separated by a few$\times 10 - 100$kpc (e.g. Arp & Crane 1992).

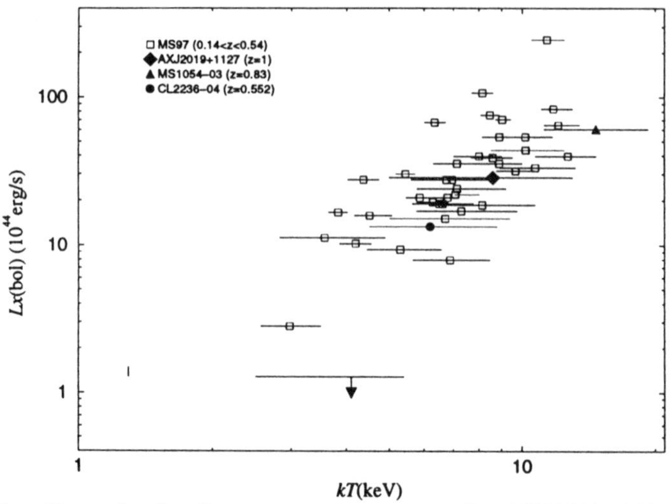

Figure 3. X-ray luminosity vs gas temperature for AXJ2019+1127 and Q2345+007 cluster. In this figure, X-ray luminosity is the bolometric luminosity where $q_0 = 0$ is assumed. The errors in the gas temperature are symmetrized 90% confidence errors. The open squares are clusters at $0.14 < z < 0.54$ from Mushotzky & Scharf (1997), filled diamond is AXJ2019+1127, filled triangle is MS1054-03 at $z = 0.83$ from Donahue et al. (1997), filled circle is CL2236-04 at $z = 0.552$ from Hattori et al. (1997b), and downward pointing arrow is 3σ upper limit of the bolometric luminosity for Q2345+007 cluster where $kT = 4.5$keV is assumed to calculate the bolometric luminosity.

References

Anders,E., Grevesse,N., (1989) *Geochimica et Cosmochimica Acta*, **53**, 197
Arnaud,M., Rothenflug,R., Boullade,O., et al., (1992) *A&A*, **254**, 49
Arp,H., Crane,P., (1992) *Phys. Lett.A*, **168**, 6
Bade,N., Siebert,J., Lopez,S., Voges,W., Reimers,D., (1997), *A&A*, **317**, 13 (BADE97)
Bonnet,H., Fort,B., Kneib,J.-P., Mellier,Y., Soucail,G., (1993) *A&A*, **280**, L5
Briel,U.G., Henry,J.P., Böhringer,H., (1992) *A&A*, **259**, L31
Chavushyan,V.H., Vlasyuk,V.V., Stcpanian,J.A., et al., (1997), *A&A*, **318**, L67 (CHA97)
Claeskens,J.-F., Surdej,J., Remy,M., (1996), *A&A*, **305**, L9 (CLAE96)
Dickey,J.M., Lockman,F.J., (1990) *ARAA*, **28**, 215
Djorgovski,S., Spinrad,H., (1984), *ApJ*, **280**, 9 (DS84)
Donahue,M., (1996) *ApJ*, **468**, 79
Donahue,M., Gioia,I., Luppino,G., Hughes,J.P., Stocke,J.T., (1997) *astro-ph/9707010*
Ebeling H., Mendes de Oliveira C., White D.A., (1995) *MNRAS*, **277**, 1006
Fischer,P., Tyson,J.A., Bernstein,G., Guhathakurta,P., (1994) *ApJ*, **431**, L71
Foltz,C.B., Weyman,R.J., Roser,H.J., Chaffee,F.H., (1984) *ApJ*, **281**, L1
Hammer,F., Le Fèvre,O., (1990), *ApJ*, **357**, 38 (HAM90)
Hammer,F., Le Fèvre,O., Angonin,M.C., et al., (1991), *A&A*, **250**, L5 (HAM91)
Hattori,M., Terasawa,N., (1993) *ApJ*, **406**, L55
Hattori,M., Ikebe,Y., Asaoka,I., Takeshima,T., Böhringer,H., Mihara,T., Neumann,D.M., Schindler,S., Tsuru,T., Tamura,T., (1997a) *Nature*, **388**, 146
Hattori,M., Matsuzawa,H., Morikawa,K., Kneib,J.-P., et al., (1997b) *in preparation*
Hattori,M., Ikebe,Y., et al., (1997c) *in preparation*

Hawkins,M.R.S., Clements,D., Fried,J.W., et al., (1997), *astro-ph/9709049* (HAW97)
Hewett,P.C., Webster,R.L., Harding,M.E., et al., (1989), *ApJ*, **346**, L61 (HEWE89)
Hewett,P.C., Irwin,M.J., Foltz,C.B., et al., (1994), *AJ*, **108**, 1534 (hewe94)
Hewett,P.C., Foltz,C.B., Harding,M.E., Lewis,G.F., (1997), *AJ*, in press (HEWE97)
Hewitt,J.N., Turner,E.L., Lawrence,C.R., et al., (1987), *ApJ*, **321**, 706 (HEWI87)
Hewitt,J.N., Turner,E.L., Lawrence,C.R., et al., (1992), *AJ*, **104**, 968 (HEWI92)
Huchra,J., Gorenstein,M., Kent,S., (1985), *AJ*, **90**,691 (HUC85)
Jauncey,D.L., et al., (1991), *Nature*, **352**, 132 (JAUN91)
Jackson,N., de Bruyn,A.G., Myers,S., et al., (1995), *MNRAS*, **274**, L25 (JAC95)
Keeton,C.R., Kochanek,C.S., (1996), *in Astrophysical Applications of Gravitational Lensing*, eds. C.S.Kochanek & J.N. Hewitt, (Kluwer: Dordrecht, Holland) 419
Kochanek,C.S., Falco,E.E., Muñoz,J.A., (1997), *astro-ph/9710165*
Langston,G.L., Schneider,D.P., Conner,S., et al., (1989), *AJ*, **97**, 1283 (LANG89)
Langston,G.I., Fischer,J., Aspin,C., (1991) *AJ.*, **102**, 1253
Lawrence,C.R., Schneider,D.P., Schmidt,M., et al., (1984), *Science*, **223**, 46 (LAW84)
Le Fèvre,O., Hammer,F., (1988), *ApJ*, **333**, L37 (LEF88)
Lehár,J., Langston,G.I., Silber,A., et al., (1993), *AJ*, **105**, 847 (LEH93)
Lehár,J., (1994), *in Gravitational Lenses in the Universe*, Ed. Surdej et al., (Liège: Université de Liège), 208 (LEH94)
Magain,P., Surdej,J., Swings,J.-P., et al., (1988), *Nature*, **334**, 327 (MAG88)
Magain,P., Surdej,J., Vanderriest,C., et al., (1992), *A&A*, **253**, L13 (MAG92)
McMahon, et al., (1992), *Gemini*, **36**, 1 (MCM92)
Mellier,Y., Dantel-Fort,M., Fort,B., Bonnet,H., (1994) *A&A*, **289**, 15
Meylan,G., Djorgovski,S., (1989), *ApJ*, **338**, L1 (MD89)
Mushotzky,R.F., Scharf,C.A., (1997) *astro-ph/9703039* (MS97)
Myers,S.T., et al., (1995), *ApJ*, **447**, L5 (MY95)
Narasimha,D., Subramanian,K., Chitre,S.M., (1987), *Astrophys. J.*, **315**, 434-439 (NSC)
Navarro,J.F., Frenk,C.S., White,S.M.D., (1995) *MNRAS*, **275**,720
Ohashi,T., (1995) *Proc. 17th Texas Symposium on Relativistic Astrophysics and Cosmology*, eds. Böhringer,H., et al., (The New York Acad. of Sciences, New York), 217
Patnaik,A.R., Browne,I.W.A., Walsh,D., et al., (1992), *MNRAS*, **259**, L1 (Pat92)
Patnaik,A.R., Browne,I.W.A., Walsh,D., et al., (1992), *MNRAS*, **259**, L9 (PAT92.2)
Patnaik,A.R., (1994), *in Gravitational Lenses in the Universe*, Ed. Surdej et al., (Liège: Université de Liège), 311 (PAT94)
Pelló,R., Miralles,J.M., Le Borgne,J.-F., Picat,J.-P., et al., (1996) *A&A*, **314**, 73
Ratnatunga,K.U., et al., (1995), *ApJ*, **453**, L5 (RAT95)
Raymond,J.C., Smith,B.W., (1977), *ApJS*, **35**, 419
Sarazin,C.L., (1986) *Rev. Mod. Phys.*, **58**, 1
Schneider,P., Ehlers,J., Falco,E.E., (1992), *Gravitational Lenses*, (Springer-Verlag)
Schneider,D.P., Lawrence,C.R., Schmidt,M., et al., (1985) *ApJ*, **294**, 66
Schneider,D.P., et al., (1986) *AJ*, **91**, 991
Simcoe,R.A., Richards,G., Turner,E.L., (1997) *private communication, in preparation*
Steidel,C.C., Sargent,W.L.W., (1991) *AJ*, **102**, 1610
Surdej,J., Magain,P., Swings,J.-P., et al., (1987), *Nature*, **329**, 695 (SURD87)
Sykes,C.M., Browne,I.W.A., Jackson,N.J., et al., (1997), *astro-ph/9710358* (SYK97)
Tinney,C.G., (1995), *MNRAS*, **277**, 609 (TIN95)
Tyson,J.A., Seitzer,P., Weymann,R.J., Foltz,C., (1986), *AJ*, **91**, 1274
van Waerbeke,L., Mellier,Y., Schneider,P., et al., (1997), *A&A*, **317**, 303
Walsh,D., Carswell,R.F., Weymann,R.J., (1979), *Nature*, **279**, 381 (WALSH79)
Weedman,D.W., Weymann,R.J., Green,R.F., et al.,(1982), *ApJ*, **255**, L5 (WEED82)
Weymann,R.J., Latham,D., Angel,J.R.P., et al., (1980), *Nature*, **285**, 641 (WEY80)
Wiklind,T., Combes,F., (1996), *Nature*, **379**, 139
Wisotzki,L., Kohler,T., Kayser,R., Reimers,D., (1993), *A&A*, **278**, L15 (WIS93)
Wisotzki,L., Köhler,T., Lopcz,S., Rcimmers,D., (1996), *A&A*, **315**, L408 (WIS96)

TABLE 2. Summary of Multiply Imaged Quasars and radio sources. From left to right, the name of source, the source redshift, the separation angle between the two main images, the magnitude of the brightest image, the type of lens where blank means yet no-detection and question mark denotes the cases either the detection of the lens is not secure or unusually high mass-to-light ratio is required, the lens redshift, the number of compact images where an E indicates the images are extended and an R indicates a ring, the degree of security as the gravitational lens where an acc denotes the acceptable case otherwise denoted by question mark, the reference for the first discovery of the source. The listed properties are mainly quoted from Keeton & Kochaneck (1996), Kochaneck, Falco, Muñoz (1997), and http: //www.stsci.edu/ftp/stsci/library/grav_lens/grav_lens.html.

Name	z_s	$\Delta\theta$ ″	m_s mag	Lens	z_l	N_{im}	GL?	Ref
LBQS2153-2056	1.85	7.8				2	?	Hewe97
Q2345 + 007	2.15	7.3	$R = 18.5$	G,CL?	1.5? 0.7?	2	?	Weed82
Q1120 + 0195	1.46	6.5	$R = 15.7$	G?	~ 0.6?	2	?	MD89
QJ0240 − 343	1.41	6.1	$B = 18.6$			2	?	Tin95
Q0957 + 561	1.41	6.1	$R = 16.7$	G,CL	0.36	2	acc	Wals79
LBQS1429-008	2.08	5.1	$R = 17.7$	$R > 24$	1.5?	2	?	Hewe89
MG0023 + 171	0.95	4.8	$r = 22.8$			2	?	Hewi87
Q2138 − 431	1.64	4.5	$B_J = 19.8$	$R > 24$		2	?	Haw97
Q1635 + 267	1.96	3.8	$B = 19.2$	G?	0.6?	2	?	DS84
MG2016 + 112	3.27	3.6	$i = 22.1$	G,CL	1.01	> 3?	acc	Law84
3C194	1.18	3.5	$R = 21.5$	G	0.31	2	?	LeF88
RXJ0911+0551	2.80	3.1	$R = 18.0$			3	acc	Bade97
HE1104 − 1805	2.32	3.1	$B = 16.9$	G?	1.66?	2	acc	Wis93
3C297	1.4	2.4	$R = 21.0$	G		2	?	Ham90
Q0142 − 100	2.72	2.2	$R = 16.8$	G	0.49	2	acc	Surd87
PG1115+080	1.72	2.2	$R = 15.8$	G,CL	0.29	4	acc	Wey80
MG0414+0534	2.64	2.1	$I' = 19.3$	G?	1.0?	4E	acc	Hewi92
Class1608+656	1.39	2.1	$R \sim 20$	G	0.63	4	acc	My95
MG1131+045	1.13?	2.1		G?	0.85?	2R	acc	Ham91
MG1654+1346	1.74	2.1	$r = 20.9$	G	0.25	R	acc	Lang89
B1938+666	< 1.0	1.8	$r = 23$	G		4R	acc	Pat94
Q2237+0305	1.69	1.7	$B = 16.8$	G	0.039	4	acc	Huc85
MG1549+3047	> 0.3?	1.7	$R = 23.3$	G	0.111	R	acc	Leh93
HE2149-2745	2.033	1.7	$B = 17.3$	G?	.2 − .5	2	acc	Wis96
SBS1520+530	1.85	1.6	$V = 18.2$			2+1?	?	Cha97
LBQS1009-0252	2.74	1.5	$B = 18.1$	G?	1.62?	2	?	Hewe94
Class1600+434	1.61	1.4	$R \sim 20$			2	acc	Jac95
HST1417+5226	3.41	1.4	$V = 24.3$	G	0.81?	4	acc	Rat95
B1422+231	3.62	1.3	$r = 15.6$	G	0.64?	4E	acc	Pat92
H1413+117	2.55	1.2	$R = 17.0$	G,CL?	1.4? 1.7?	4	acc	Mag88
HST1253-2914		1.2	$V = 25.5$	G		4	acc	Rat95
Class1933+503		1.1	$I > 24.5$	G	0.755	4+5	acc	Syk97
PKS1830-211		1.0		G	0.89	2ER	acc	Jaun91
BRI0952-0115	4.5	0.95				2	?	McM92
MG0751+2716		0.9		G	0.35?	4R	acc	Leh94
J03.13	2.55	0.84	$R = 17.1$			2	acc	Clae96
Q1208+1011	3.80	0.48	$V = 18.1$			2	?	Mag92
B0218+357	0.96?	0.34		G	0.68	2ER	acc	Pat92.2

ARE ELLIPTICAL GALAXIES REALLY METAL-RICH?

And, If Not, Then What?

M. LOEWENSTEIN AND R. F. MUSHOTZKY
NASA/GSFC
Code 662, Greenbelt, MD 20771, USA

1. Introduction

In elliptical galaxies, where most of the stars – and therefore most of the heavy elements – were formed at an early epoch, the total mass, spatial distribution, and relative abundances of metals are intimately connected to the galaxy formation process.

Metallicities of the stars in elliptical galaxies can be estimated from optical broad-band photometry and spectroscopic measurement of selected absorption line indices. However, the transformation from these indirect abundance indicators to true stellar metallicites is highly non-trivial, depending on *a priori* assumptions about stellar physics details and star formation history. Since the hot interstellar medium primarily originates from stellar mass loss, X-ray observations provide independent and complementary abundance information. Moreover, there are distinct advantages to using strong X-ray emission features over weak optical absorption features: the former are related to actual abundances in a much more direct, model-independent, manner; and, one can determine abundances much further out in radius. 30-40 early-type galaxies have been observed with *ASCA*, representing the first large sample of high quality, broad-band X-ray spectra of elliptical galaxies. Most of these have been analyzed in detail by K. Matsushita, and we draw upon the results presented in her thesis, as well as our own analysis, in this review.

2. Abundances in Gas-Rich Ellipticals

ASCA spectra can generally be decomposed into soft and hard components (Matsumoto *et al.* 1997). The soft component originates in the hot (0.3-1 keV) ISM, and shows a wide range of X-ray-to-optical flux ratios and X-ray extents for any given optical luminosity. The hard component generally

K. Nomoto and J.W. Truran (eds.), Cosmic Chemical Evolution, 139–146.

scales linearly with optical luminosity, with a relative normalization and spectrum consistent with measurements of the integrated emission from low mass X-ray binaries in spiral galaxy bulges. Figure 1 (from Matsumoto *et al.* 1997) compares *ASCA* Solid State Imaging Spectrometer (SIS) spectra of the elliptical galaxies NGC 4365 and NGC 4636. Note that these galaxies have similar optical luminosities and hard components; however, the soft component in NGC 4636 is nearly two orders of magnitude more luminous than that in NGC 4365. Since abundance uncertainties become large as the hard component begins to dominate and the equivalent widths of emission lines are diluted (e.g., for NGC 4365), this review focuses on observations of eighteen gas-rich ellipticals (e.g., NGC 4636) in the *ASCA* archive – galaxies with high X-ray-to-optical fluxes based on *ROSAT* PSPC observations and low hard component fractions based on *ASCA* spectral decomposition. The X-ray emission extends well beyond the optical isophotes for some of these systems, but is more compact in others.

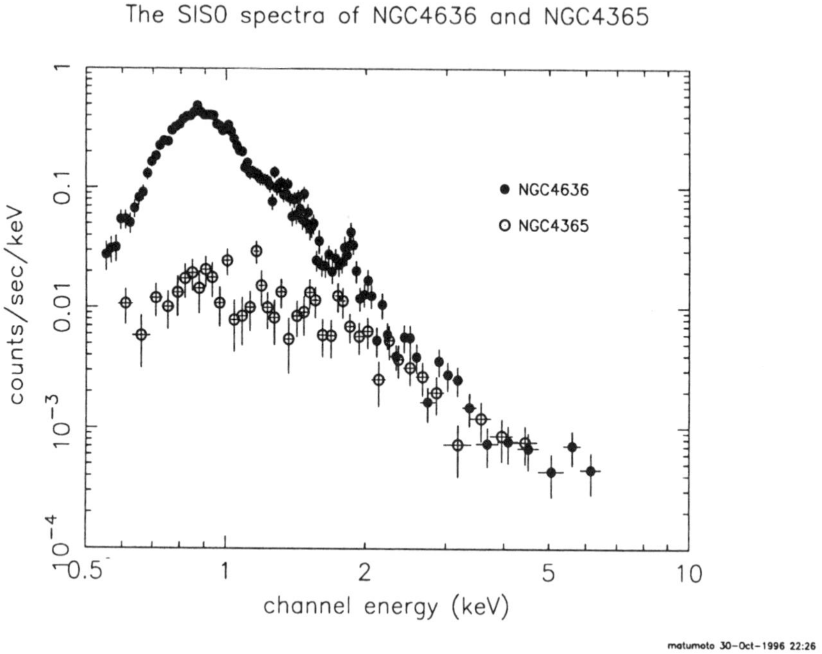

Figure 1. Co-added *ASCA* SIS spectra of NGC 4636 (filled circles) and NGC 4365 (open circles).

Figure 2 shows a plot of abundance versus temperature derived from *ASCA* spectra extracted from the inner five optical half-light radii. The soft component is modeled using the Raymond-Smith thermal plasma emission code with abundances fixed at their solar photospheric ratios. The abundances – essentially the Fe abundance, since X-ray spectra at these temperatures are dominated by Fe L emission lines – range from about one-tenth to two-thirds solar. Since it has generally been assumed that abundances of the mass-losing stars that are the origin of the hot gas are supersolar, and that Type Ia SN should further enrich the hot gas by at least an additional two times solar, X-ray abundances of elliptical galaxies are 3-30 times lower than what might naively be expected.

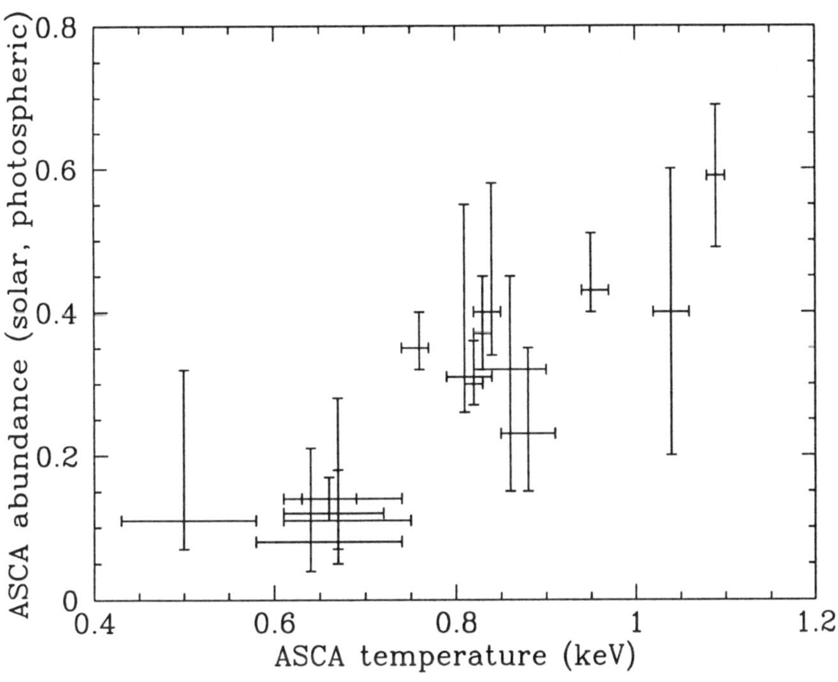

Figure 2. Hot gas metal abundance versus temperature.

3. Reconciling Optical and X-ray Abundances

Clearly, either the SNIa rate is much lower than estimated, and/or SNIa ejecta is not efficiently mixed into the hot ISM. Can optical and X-ray abundances be reconciled, even if SNIa enrichment is neglected? It must be the case that either (1) stellar abundances are not actually supersolar, (2) hot gas abundances are not actually subsolar, or (3) X-ray and optical observations are not commensurate. We consider these possibilities in reverse order.

(3) We now know that the X-ray emission from some elliptical galaxies is much more extended than the optical light and, accounting for the measured abundance distributions, that the radius containing half of the hot gas metals is 3-10 times larger than the corresponding radius for the stars in such systems. The hot gas metallicity profile generally has a negative gradient (Matsushita 1997). These facts indicate that accretion of primordial or intergalactic material may be diluting the abundances of the hot gas relative to the stars (Brighenti & Mathews 1997). However, if this were a dominant effect one would expect lower average metallicities in more gas-rich galaxies – if anything, the opposite is observed. Although gas flows have undoubtedly rearranged and in some cases diluted the gas, it is probably meaningful to compare optical and X-ray metallicities within the optical radius.

(2) The validity of the abundances obtained from single-phase hot gas plus X-ray binaries fits to *ASCA* data have been questioned on a number of fronts. Inadequacies in the treatment of Fe L transitions in plasma codes used for spectral fitting are an extra source of uncertainty (Arimoto *et al.* 1997); however, excluding the energy region in question in spectral fits does not systematically raise the metallicity (Buote & Fabian 1997). Higher abundances can also be accommodated in more complex spectral models. In particular, Buote and Fabian have recently found that the best fit to *ASCA* data often consists of a two-temperature plasma, with the secondary component having a temperature of $\sim$ 1.5 keV, and that the abundances in such fits are systematically higher by about a factor of two compared to models with a single gas phase plus X-ray binaries. However, there is more information in *ASCA* spectra than what can be obtained from global spectral fits alone. We have found that the temperature obtained from the He-like to H-like Si line ratios is in excellent agreement with the single-phase model (see Figure 3 for an example with a typical level of agreement), and is not consistent with the presence of hotter gas in the amounts suggested by Buote and Fabian.

(1) So, it seems reasonable to adopt the metallicities in the single-phase hot gas plus X-ray binaries fits to *ASCA* spectra as measures of global interstellar Fe abundances, and sensible to compare these with global Fe abundances of the stars. What are the best current estimates of the latter? The conventional wisdom that elliptical galaxies have supersolar abundances is based on measurements of the nuclear Mg2 index. However, when one accounts for the factor of 2-3 overabundance of Mg with respect to Fe in the stars (Worthey *et al.* 1992), and the factor of two aperture correction (Arimoto *et al.* 1997) it becomes clear that the global Fe abundances of the stars and hot gas are not grossly discordant. We have compared recent estimates of the global Fe abundance using optical data kindly provided by S. Trager

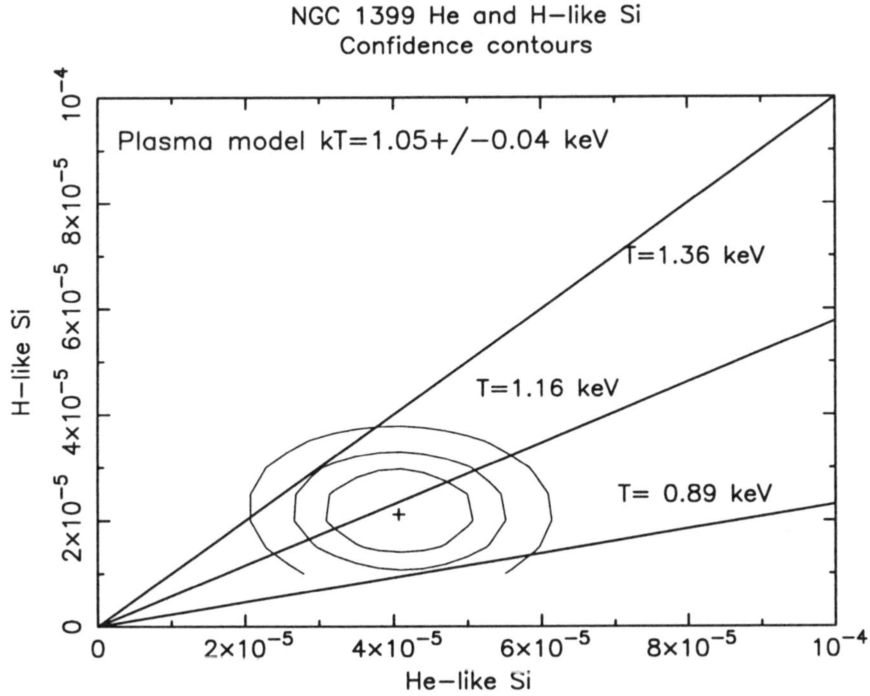

Figure 3. 68, 90, and 99% confidence contours for H- and He-like Si line strengths (in photons cm^{-2} s^{-1}) in the elliptical galaxy NGC 1399. The solid lines show the ratios expected in three single-temperature thermal plasma models. The best-fit temperature to the spectrum derived from global Raymond-Smith plasma model fits is 1.05 ± 0.04 keV (90% confidence uncertainties).

(Trager 1997) with the X-ray measurements. The comparison for the eight galaxies present in both samples is shown in Figure 4. The discrepancy is generally minor, although a few galaxies still have unaccountably low X-ray abundances. The average optical Fe abundances are about 0.45 solar (as predicted in the chemically consistent evolutionary models of Moller *et al.* 1997) compared to 0.3 solar for the hot gas.

4. Relative Abundances

Elemental abundance ratios provide constraints on the primordial IMF and relative numbers of Type Ia and Type II supernovae. Given sufficient signal-to-noise, O, Mg, Si, S, and Fe all have prominent, distinct features in *ASCA* spectra of elliptical galaxies. In variable abundance fits of seven ellipticals, we find Si-to-Fe ratios consistent with or perhaps somewhat less than the (meteoritic) solar value (Figure 5). This is lower than the Mg-to-Fe ratio derived from nuclear optical spectra, and is more in line with what has been

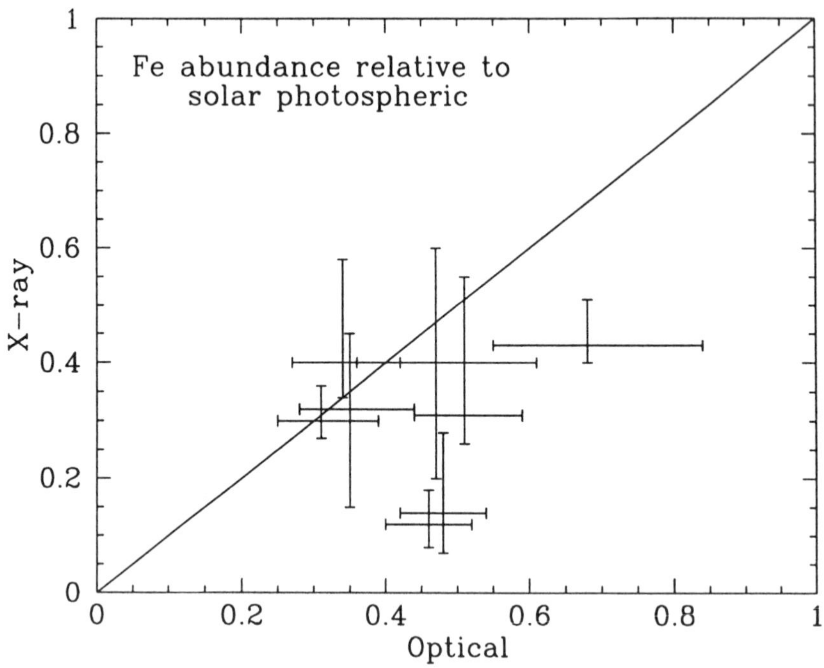

Figure 4. X-ray versus optical global iron abundance.

measured from the X-ray spectra of intergroup media. The Si abundance provides an independent and robust (plasma code uncertainties are very small for these Si lines) strong upper limit on the effective SNIa rate that is consistent with what is derived using Fe. The limit of about 0.03 SNU (1 SNU = 1 supernova per century per $10^{10} L_{B\odot}$) is about four times lower than the recent estimate of Cappellaro *et al.* (1997) for $H_o = 75$ km s^{-1} Mpc^{-1}

5. Implications of Low Abundances for ICM Enrichment

We have shown that the mass-averaged Fe abundance in an elliptical galaxy, based on X-ray and optical observations, is about one-half solar. This is only slightly higher than what is typically measured in the intracluster medium, and yet the mass in the ICM is generally 2-10 times greater. There is several times more Fe (and Si, as well) in the ICM than is locked up in stars in cluster galaxies. This implies the following.

(1) If the stellar and ICM metals come from the same SNII-enriched proto-elliptical galaxy gas, then 50-90% of the original galaxy mass has been lost and a significant fraction of the ICM is not primordial but has

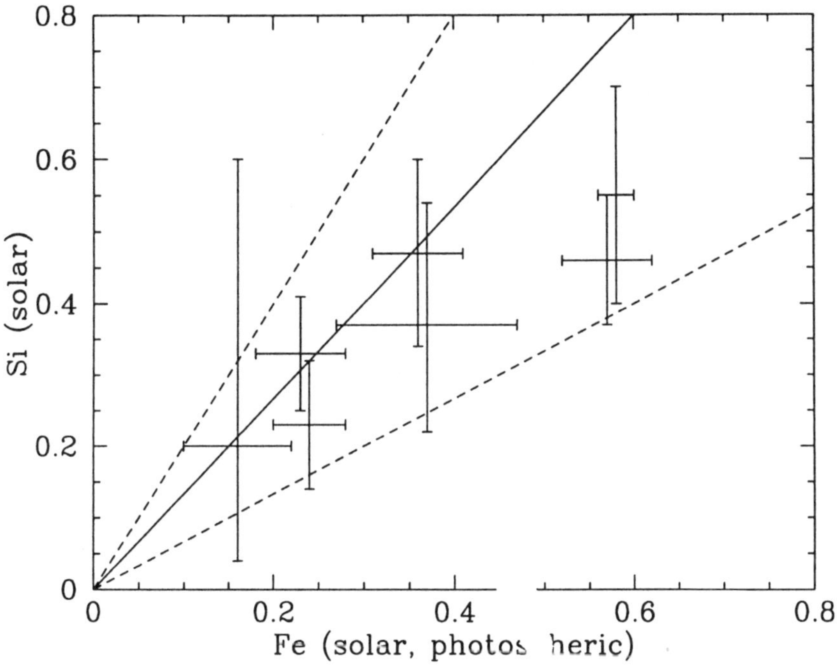

Figure 5. Si versus Fe abundance in the hot X-ray emitting gas. The solid line denotes Si:Fe in the ratio 1:1, while the broken lines denote the ratios 3:2 and 1:2 with respect to the (meteoritic) solar ratio.

been ejected from galaxies. (2) However, the actual amount of material directly associated with the SNII ejecta is roughly an order of magnitude less. If there is selective mass-loss of nearly pure SNII ejecta it is possible to lose most of the metals without losing most of the mass.

(3) It is also possible that there is an additional significant source of ICM enrichment, although there are some difficulties with such a scenario (Gibson and Matteucci 1997). Perhaps, alternate sources of enrichment should be re-examined in light of the downward revision of elliptical galaxy abundances; although, one must bear in mind that ICM Fe mass is highly correlated with the total luminosity in elliptical galaxies (Arnaud *et al.* 1992).

6. Concluding Remarks

X-ray spectra of elliptical galaxies are adequately fit by models consisting of hot gas with subsolar Fe abundance and roughly solar Si-to-Fe ratio, plus a hard component from an ensemble of X-ray binaries. The consistency of the relative magnitude and spectrum of the hard component with that expected from X-ray binaries, along with its more compact spatial distribution supports this model over ones where the hard component is primarily due to a hotter gas phase. Complications in the form of an extra soft continuum or multiple phases can be considered, but the consistency of the Si line diagnostic and continuum temperatures demonstrates that the data – at the present level of sensitivity and spectral resolution – do not require these. Optical and X-ray Fe abundance estimates are converging, although there are some cases with anomalously low X-ray values. Problems in the Fe L spectral region remain; however, the main effect of improvements in atomic physics parameters is likely to be improved spectral fits rather than a radical upward revision in abundances.

Occam's razor would seem to demand that we provisionally accept the reality of low abundances in elliptical galaxies. As a result, we need to seriously reevaluate our notions of elliptical galaxy chemical evolution, intracluster enrichment, and Type Ia supernova rates.

We are grateful to Scott Trager and Kyoko Matsushita for making results from their dissertations available to us prior to publication.

References

Arimoto, N., Matsushita, K., Ishimaru, Y., Ohashi, T., & Renzini, A. 1997, *ApJ*, **477**, 128

Arnaud, M., Rothenflug, R., Boulade, O., Vigroux, L., & Vangioni-Flan, E. 1992, *A& A*, **254**, 49

Brighenti, F. & Mathews, W. G. 1997, *ApJ*, in press

Buote, D. A., & Fabian, A. C. 1997, *MNRAS*, submitted

Cappellaro, E., Turatto, M., Tsvetkov, D. Yu., Bartunov, O. S., Pollas, C., Evans, R., & Hamuy, M. 1997, *A& A*, **322**, 431

Gibson. B. K., & Matteucci, F. 1997, *ApJ*, **475**, 47

Matsumoto, H., Koyama, K., Awaki, H., Tsuru, T., Loewenstein, M., & Matsushita, K. 1997, *ApJ*, **482**, 133

Matsushita, K. 1997, Ph.D. thesis, University of Tokyo

Moller, C. S., Fritze-v. Alvensleben, U. & Fricke, K. J. 1997, *A& A*, **317**, 676

Trager, S. C. 1997, Ph.D. thesis, University of California, Santa Cruz

Worthey, G., Faber, S. M., & Gonzalez, J. J. 1992, *ApJ*, **398**, 69

CHEMICAL EVOLUTION OF SPIRAL GALAXIES FROM REDSHIFT 4 TO THE PRESENT

U. FRITZE - V. ALVENSLEBEN, U. LINDNER, K. J. FRICKE
Universitäts–Sternwarte, Göttingen, Germany

1. Introduction

ISM abundances in nearby spiral galaxies are well known from HII region studies (Zaritsky *et al.* 1994). While early type spirals, Sa, Sb, have rather uniform abundances and a narrow range of present star formation rates (**SFR**) the galaxy-to-galaxy variations both in HII region abundances and in present SFR increase towards late spiral types Sc, Sd (see e.g. Kennicutt & Kent 1983). ISM abundances of spiral galaxies or their progenitors up to the highest redshifts can be studied via the absorption properties imprinted in the spectra of background QSOs. While MgII- and CIV- absorption lines are produced in the low column density gas of the extended haloes around galaxies, the Damped Lyα Absorption (**DLA**) is believed to originate in (proto-)galactic disks. High resolution spectroscopy of a large number of metal lines associated with DLA systems reveal the redshift evolution of ISM abundances from z $\gtrsim$ 4 to z $\sim$ 0.6.
We present a chemo-cosmological evolution model to explore a possible link between high redshift DLA systems and nearby spirals. Inclusion of a spectrophotometric description in parallel to the chemical modelling together with the enormous time baseline corresponding to the redshift range until z $\sim$ 4 allow the galaxy parameters to be seriously constrained.

2. Galaxy Models

Any kind of evolutionary galaxy model requires two fundamental parameters: the star formation history (**SFH**) and the initial mass function (**IMF**). For a given set of input physics – stellar evolutionary tracks, yields, spectra, ... – these two parameters govern both the spectral and the chemical

K. Nomoto and J.W. Truran (eds.), Cosmic Chemical Evolution, 147–151.

evolution of a model galaxy.
Our unified chemical and spectrophotometric galaxy evolution model describes in detail the spectrophotometric evolution in terms of luminosities $L_U, \ldots, L_K$, colours, gaseous emission lines, stellar absorption indices, and the chemical evolution in terms of ISM abundances of individual elements ^{12}C, ... , ^{56}Fe, including SNI contributions from carbon deflagration white dwarf binaries á la Matteucci (1991). All these quantities are obtained as a function of time, and, for a given cosmological model – characterised by H_0, Ω_0, Λ_0 and a galaxy formation redshift z_{form} – in terms of redshift z.

A specific virtue of this kind of evolutionary model is its analytic power. Luminosities at any wavelength and the enrichment of any element are readily decomposed into contributions from different stellar masses, spectral types, luminosity classes, metallicity subpopulations, nucleosynthetic sites (PNe, SNI, SNII, single stars, binaries, ...), and all this can be followed as a function of time or redshift. It's a simplified 1 – zone model without any dynamics or spatial resolution but it consistently accounts for the finite lifetime of each star, i.e. it does not use an Instantaneous Recycling Approximation.
To cope with the low metallicities of DLAs we have recently extended our code to consistently account for the increasing metallicities of successive generations of stars by using sets of input physics (stellar tracks, yields, spectra, ...) for a range of metallicities from $Z = 0$ to $Z = 2 \cdot Z_\odot$. Sources for the stellar yields for various metallicities are v. d. Hoek & Groenewegen (1997) for stars with masses $m_* \leq 8\ M_\odot$ and Woosley & Weaver (1995) for stars in the range $12 \leq m_* \leq 40\ M_\odot$.
We use different SFHs for the different galaxy types in order to assure detailed agreement of our model galaxy spectra with template spectra for the respective spectral class from Kennicutt (1992). While ellipticals and haloes are well described with a SFR exponentially decreasing in time, spirals are more adequately described by SFRs that are linear functions of the evolving gas content G(t). Characteristic timescales for SF t_*, as defined by the decrease rate of the initial gas content G_0: $\int_0^{t_*} \Psi \cdot dt = 0.63 \cdot G_0$ are $t_* \sim 1$ Gyr (Es & Halos), $t_* \sim 2$ Gyr (Sa), ~ 3 Gyr (Sb), ~ 10 Gyr (Sc), ~ 16 Gyr (Sd).
While spectrophotometric modelling in comparison to observed galaxy spectra and colours allows for reasonable determinations of the SFHs of galaxies of various spectral types it gives only weak constraints on the IMF. Detailed chemical modelling, on the other hand, constrains both the SFHs **and** the IMF through comparison with observations of the redshift evolution of abundances and abundance ratios of elements originating from different nucleosynthetic sites, as e.g. [C/O], [O/Fe], or [Mg/Fe], for an enormous time baseline. The lookback time to redshifts $z \sim 4$ corresponds

to $\gtrsim 90\%$ of the Hubble time.
Fritze - v. Alvensleben *et al.* (1989, 1991) model the redshift evolution of halo abundances and compare with MgII- and CIV- data. HST observations of CIV at low redshift (Bahcall *et al.* 1993) confirm our prediction for carbon.

3. Redshift Evolution of DLA Abundances

Damped Lyα Absorbers usually feature a series of associated low ionisation absorption lines, e.g. of C, N, O, Al, Si, S, Cr, Mn, Fe, Ni, Zn, ...
High resolution spectra allow to derive precise element abundances, which are, by now, available for a large number of DLAs over the redshift range from $z \sim 0.6$ through $z \gtrsim 4.4$ (Lu *et al.* 1996, Pettini *et al.* 1997). Considerable care is taken to resolve the velocity structure, to only use non- (or weakly) saturated lines and dominant ionisation stages, and to a possible depletion onto dust grains that may vary dramatically from ~ 0 % (for Zn) to $\gtrsim 90$ % (for Cr). We have compiled all available DLA abundances and have carefully referred them all to one "standard" set of oscillator strengths and solar reference values. Observational abundances of DLAs show a weak increase with decreasing redshift – in particular if compared to CIV - halo - systems – and a large scatter at any redshift.

4. Comparison of DLA Abundances with Models for $Z_\odot$

For our modelling approach it would not matter if indeed they highest redshift DLA systems were a bunch of subgalactic fragments rather than one single protodisk. As long as the ensemble of fragments is bound to merge into the counterpart of a present-day galactic disk our model SFH is then meant to describe the SFH of the ensemble.
Using solar metallicity stellar yields, lifetimes and evolutionary tracks as input physics for our modelling, a Scalo IMF, and a "standard cosmology" ($H_0 = 50, \Omega_0 = 1, \Lambda_0 = 0, z_{form} = 5$), we find the following results (Fig. 1):

- Sa – Sd models nicely bracket the DLA abundance data, i.e.,
- for SF timescales t_* typical of spiral galaxies we find an evolution of abundances (and element ratios) which is in good agreement with observations of DLAs over a large portion of the Hubble time,
- the models provide a smooth transition from DLAs to nearby spirals' HII region abundances (Zaritsky *et al.* 1994),
- the range of SF timescales t_* as empirically derived for the near-by Hubble sequence of spiral galaxies from Sa through Sd fully explains the observed scatter of metallicities among DLAs at fixed redshift.

(cf. Fritze - v. Alvensleben & Fricke 1995, Fritze - v. Alvensleben 1995).

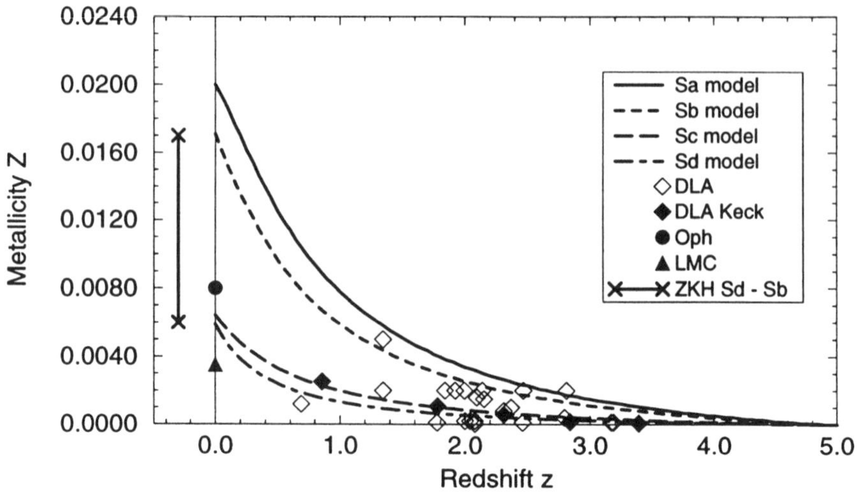

Figure 1. Redshift evolution of the global metallicity: spiral models vs. DLA. ZKH: range of HII region abundances in nearby spirals of types Sb through Sd (Zaritsky *et al.*). $(H_0, \Omega_0, z_{form}) = (50, 1.0, 5)$.

We conclude: Our chemical evolution models confirm the hypothesis that DLA systems could be the progenitors of present day spirals and predict early as well as late type spirals to be among the DLA sample.
In a complementary investigation Lindner *et al.* (1996, 1998) compare the spectrophotometric results from our unified chemical **and** spectrophotometric model to a large sample of optically identified QSO absorbers.

5. Results from our Chemically Consistent Modelling

In view of the low metallicities of DLAs a chemically consistent description is clearly desired. A problem with this approach is that important pieces of input physics, as e.g. stellar yields, explosion energies, lifetimes, mass loss rates, the final fate of a star and its remnant mass for very low metallicities are far from being fully understood. In this sense, the first results we now present from our chemically consistent modelling deserve some caution.

Due to the complicated behaviour of the yields for individual elements as a function of stellar mass and metallicity, changes in the redshift evolution of chemically consistent models with respect to solar metallicity models vary for different elements, and, of course, also with galaxy type (=SFH).

For Zn, the most common metallicity tracer in DLAs, our Sa – Sd models nicely bracket recent and precise DLA abundance data over the redshift range $0.6 \lesssim z \lesssim 3.5$ as seen in Fig. 2.

Abundance ratios change drastically as compared to solar metallicity models. This has far-reaching consequences e.g. for the interpretation of the

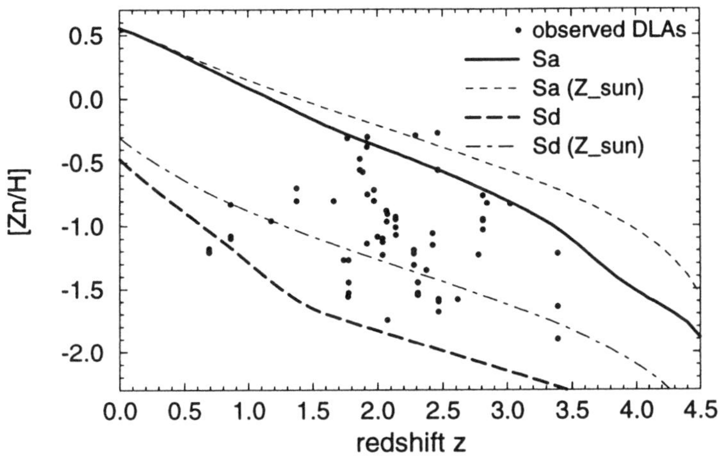

Figure 2. Redshift evolution of [Zn/H] for spiral galaxy models together with DLA data.

stellar [Mg/Fe] in elliptical galaxies (see Fritze - v. Alvensleben 1998).

It is clearly desirable to extend this study to other elements with high resolution data. Our model offers a unique possibility to study the redshift evolution of the Global Cosmic SFR in the Universe including consistently all three observational accesses: gas content, chemical enrichment and spectral properties.

UFvA greatfully acknowledges a travel grant (Fr 916/4-1) from the DFG.

References

Bahcall, J. N. *et al.* , 1993, ApJS **81**, 1

Fritze - v. Alvensleben, U., 1995, in *QSO Absorption Lines*, ESO Workshop, p. 81

Fritze - v. Alvensleben, U., 1998, in *Abundance Ratios of the Oldest Stars*, Highlights of Astronomy, *in press*

Fritze - v. Alvensleben, Fricke, K. J., 1995, IAU Symp. **164**, 457

Fritze - v. Alvensleben, U., Krüger, H., Fricke, K. J., Loose, H.-H., 1989, A&A **224**, L1

Fritze - v. Alvensleben, U., Krüger, H., Fricke, K. J., 1989, A&A **246**, L59

Kennicutt, R. C., 1992, ApJS **79**, 255

Kennicutt, R. C., Kent, S. M., 1983, ApJ **88**, 1094

Matteucci, F., 1991, ASP Conf. Ser. **20**, 539

Lindner, U., Fritze - v. Alvensleben, Fricke, K. J., 1996, A&A **316**, 123

Lindner, U., Fritze - v. Alvensleben, Fricke, K. J., 1997, in *Structure and Evolution of the IGM from QSO Absorption Line Systems*, ed. Petitjean, *in press*

Lindner, U., Fritze - v. Alvensleben, Fricke, K. J., 1998, *this volume*

Lu, L., Sargent, W. L. W., Barlow, et al., 1996, ApJS **107**, 475

Pettini, M., Smith, L. J., King, D. L., Hunstead, R. W., 1997, ApJ **486**, 665

Stengler-Larrea, E., 1995, in *QSO Absorption Lines*, ESO Workshop, p. 199

van den Hoek, L. B., Groenewegen, M. A. T., A&AS **123**, 305

Woosley, S. E., Weaver, T. A., 1995, ApJS **101**, 181

Zaritsky, D., Kennicutt, R. C., Huchra, J. P., 1994, ApJ **420**, 87

AGE PROBLEM BASED ON THE ABUNDANCE RATIO FROM A HIGH-REDSHIFT QSO

T. TSUJIMOTO
National Astronomical Observatory, Mitaka, Tokyo 181-8588, Japan

AND

Y. YOSHII, K. KAWARA
Institute of Astronomy, Faculty of Science, University of Tokyo, Mitaka, Tokyo 181-8588, Japan

Abstract. The age of a high-redshift QSO can constrain the cosmology by placing a limit on the cosmological time. We use the observed $\mathrm{Fe_{II}}$ (UV+optical)/$\mathrm{Mg_{II}}\lambda 2798$ flux ratio from the broad-line gas of a QSO at z=3.62 to estimate the abundance ratio [Mg/Fe] and predict that the QSO is at least 1.3 Gyr old from the chemical analysis using the abundance ratio [Mg/Fe] as a cosmic clock. Our result favors a nonzero cosmological constant and thus conflicts with a flat, matter-dominated universe.

1. Introduction

Type II supernovae (SNe II) contribute to the heavy element production in the early chemical evolution of galaxies. Type Ia supernovae (SNe Ia), in contrast, produce heavy elements with a considerably longer timescale in the late phases of the galactic evolution, as they are due to thermonuclear explosions of accreting white dwarfs. The yields of SNe II and SNe Ia and thus their nucleosynthesis roles in the chemical evolution of galaxies are distinctly different, i.e., SNe Ia eject much larger amount of Fe than SNe II, but have negligible contributions to the O and Mg abundances (see Tsujimoto et al. 1995). These differencies between SNe II and SNe Ia enable us to measure the age of galaxies (QSOs) by using the abundance ratios of O/Fe or Mg/Fe as cosmic clocks.

Here we determine the lower limit on an age of QSO B1422+231 at z=3.62 from the abundance ratio [Mg/Fe], which is estimated from the observed flux ratio $\mathrm{Fe_{II}}$(UV + optical)]/$\mathrm{Mg_{II}}\lambda 2798$ obtained by Kawara et

K. Nomoto and J.W. Truran (eds.), Cosmic Chemical Evolution, 153–157.

al. (1996). The derived age puts the powerful constraint on cosmology, suggesting that there exits the so-called age problem at z=3.62 as well as at z=0 and 1.55.

2. Age Problems at z=0, 1.55

In this section we mention the present status of the classical age problem based on the Galactic globular cluster age and another age problem resulting from a recent discovery of a 3.5-Gyr-old galaxy at z=1.55.

2.1. AGE PROBLEM AT Z=0

Ages of Galactic globular clusters (GCs) give the lower limit on the age of the universe. The age of a GC can be deduced from stellar evolution calculations and is estimated to be about 15 Gyr (e.g., 15.8±2.1 Gyr by Bolte & Hogan 1995; 14.56±2.49 Gyr by Chaboyer et al. 1996). On the other hand, the standard cosmological model of a flat, matter-dominated universe with the Hubble constant H_0= 60–80 km s^{-1} Mpc^{-1} estimated from almost all observations implies a cosmic age of 8–11 Gyr, which is inconsistent with the ages of Galactic GCs. To solve this discrepancy, we should adopt the density of the universe $\Omega < 0.2$, or introduce the cosmological constant Λ if the estimated ages of GCs are correct. Recently, a possibility of younger age of GCs has been discussed since the preference of the bright absolute magnitude of RR Lyrae stars M_V(RR) was inferred from the Hipparcos measurements of parallaxes of Cepheids (Feast & Catchpole 1997) and subdwarfs (Reid 1997). However, Tsujimoto, Miyamoto, & Yoshii (1997) present the *direct* determination of M_V(RR) from statistical treatments of Hipparcos proper motions and trigonometric parallaxes of RR Lyraes and have confirmed the faint M_V(RR), thus suggesting that the so-called age problem would not be resolved or alleviated yet.

Moreover, a simple, independent estimate of the age of one very old star (CS22892-052), using a method called nucleocosmochronology, is presented by Sneden et al. (1996). From the deficiency in a radio-active r-process element Th, they derive the age of 15.2±3.7 Gyr, which is in good agreement with the ages of GCs.

2.2. AGE PROBLEM AT Z=1.55

Dunlop et al. (1996) discoverd a red radio galaxy LBDS 53W091 at z=1.55. By fitting the observed spectrum to stellar population models and taking advantage of the age-sensitivity of the near-ultraviolet spectrum, Spinrad et al. (1997) derive a minimum age of 3.5 Gyr. On the other hand, the Universe is only 2–2.6 Gyr old at z=1.55 for a flat, matter-dominated universe. This is the second age problem.

3. Age Problem at z=3.62

The rest-frame UV-optical spectrum of the gravitational lensed QSOB1422 +231 at z=3.62 is presented by Kawara et al. (1996). Their observed Fe_{II}(UV+optical) /$Mg_{II}\lambda 2798$ flux ratio is 12.2±3.9. Similar flux ratios $\sim$ 8 are obtained for QSOs at low and intermediate redshifts (Wills, Netzer, & Wills 1985). Wills et al. also calculate the strengths of the emission lines by their photoionization models and conclude that the observed flux ratio $\sim$ 8 suggests an overabundance of Fe by a factor of $\sim$ 3 relative to Mg. Therefore the lower bound for the observed flux ratio from B1422+231 corresponds to the abundance ratio [Mg/Fe] ~ -0.5. It means that B1422+231 has the abundance ratio [Mg/Fe] $\lesssim$ -0.5.

It is noted that the abundance ratio [Mg/Fe] in B1422+231 is quite different from those in the solar neighborhood stars. The negative ratio implies a significant contribution of Fe from SNe Ia. Such a large contribution from SNe Ia to the broad-line gas is possible if QSOs are associated with vigorous star formation. This effect is understood as follows. The heavy elements ejected from SNe II during the burst are largely trapped in stars and a small fraction of the heavy elements is contained in the gas. On the other hand, SNe Ia start to occur after the burst, then the heavy elements ejected from SNe Ia are restored in the gas because of the absence of star formation. Therefore after the burst, the gas becomes considerably enriched with Fe from SNe Ia.

Hamann & Ferland (1992,1993) construct the chemical evolution models for QSOs so as to explain the N_V/C_{IV} line ratios at high-z QSOs. They conclude that high-z QSOs were enriched with heavy elements in a burst of star formation, which is consistent with an above implication of [Mg/Fe] in B1422+231. They derive enhanced star formation rates for relatively low-metallicty QSOs (ν=5.0), high-metallicty QSOs (ν=6.7), and extreme high-metallicity QSOs (ν=7.6), respectively [ν: the coefficient of star formation rate; ν=0.27 for their solar neighborhood model]. The star formation rate for extreme high-metallicity QSOs gives the shortest timescale for the enrichment by heavy elements in QSOs.

The lifetime of SN Ia progenitors as well as the star formation rate is a critical parameter for calculating the time evolution of Fe. The lifetime t_{Ia} of SN Ia progenitors is deduced from the chemical evolution in the solar neighborhood (Yoshii, Tsujimoto, & Nomoto 1996). It is found that the range of t_{Ia} is strictly confined within 0.5–3 Gyr.

Using the above determined parameters together with a Salpeter initial

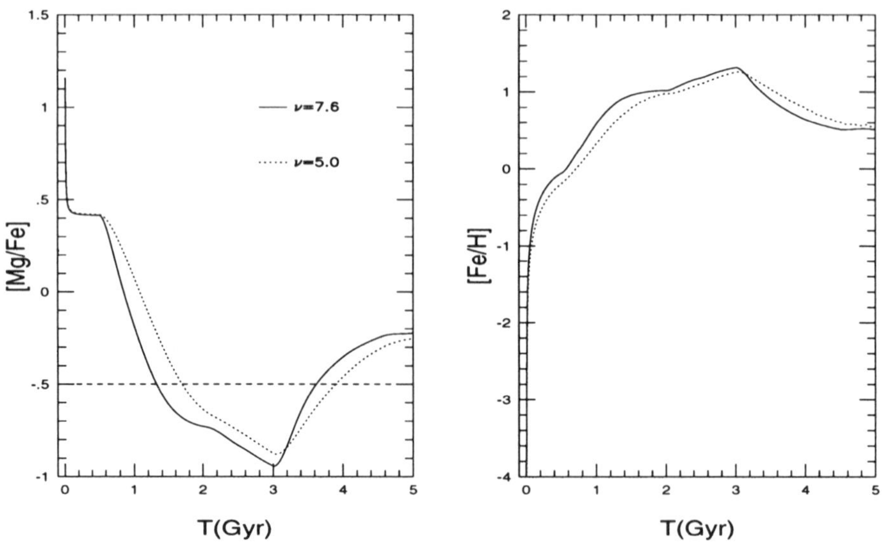

Figure 1. The evolutionary change in [Mg/Fe] and [Fe/H] predicted by the models with high star formation rates.

mass function (IMF), we calculate the evolutionary change in [Mg/Fe] and [Fe/H] in QSOs (Figure 1). The dashed line denotes the upper bound for the observed [Mg/Fe]. The intersection between the theoretical curve of [Mg/Fe] predicted by the model with the extreme high star formation rate and the observed upper bound corresponds to the strong lower limit on the age of the QSO B1422+231. We therefore obtain $T_{\rm QSO\ at\ z=3.62} \geq 1.3$Gyr, whereas the Universe is only 0.6–0.8 Gyr old at z=3.62 for a flat, matter-dominated universe.

Figure 2 shows the predicted age–redshift relations for three cosmological models, i.e., a flat, matter-dominated universe (Ω=1, Λ=0), an open universe (Ω=0.2, Λ=0), and a cosmological constant-dominated universe (Ω=0.2, Λ=0.8), together with three age constraints. The solid lines for each panel denote the cases for H_0= 60 and 80 km s^{-1} Mpc^{-1}, assuming the formation epoch at z=10. The case for the formation epoch at z=5 with H_0= 60 km s^{-1} Mpc^{-1} is also shown by the dashed line.

4. Conclusion

A flat, matter-dominated universe conflicts with age problems at three different redshifts, i.e., z=0, 1.55 and 3.62. The age constraint from a high-z QSO is more stringent than those from the Galactic globular cluster or a galaxy at z=1.55. An open universe ($\Omega = 0.2$) is acceptable if we adopt H_0

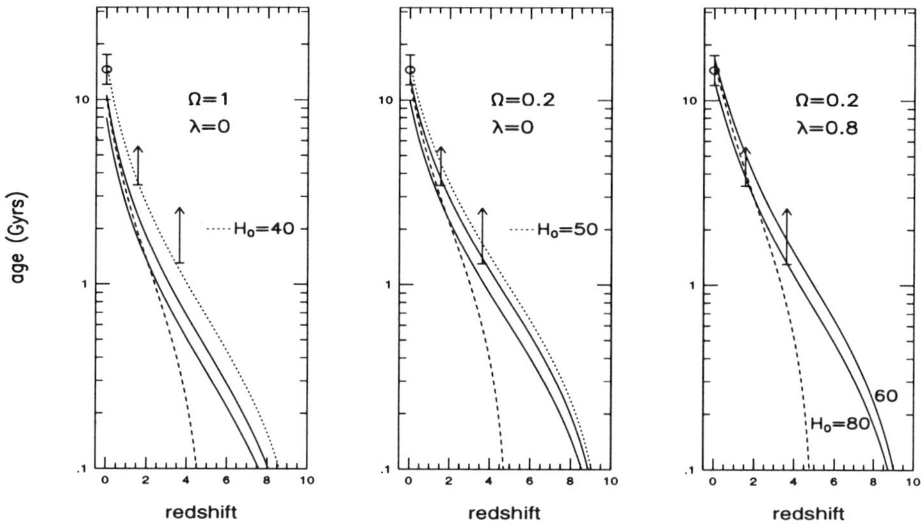

Figure 2. The age–redshift relations for three cosmological models.

less than 65 km s^{-1} Mpc^{-1}. However a preferable age constraint at z=3.62 inferred from the observed Fe_{II}(UV+optical) /$Mg_{II}\lambda2798$ flux ratio $\sim$ 12 excludes a open universe. This age problem is more serious if the IMF in QSOs is weighted toward forming more massive stars than in the solar neighborhood. It is likely that in the starburst the IMF has a shallower slope or has a truncation at a large lower stellar mass limit. In this case the calculated [Mg/Fe] becomes higher and it results in giving the older age for the QSO. On the other hand, this age constraint is consistent with the cosmological time predicted by a cosmological constant-dominated universe.

References

Bolte, M. and Hogan, C.J. 1995, *Nature*, **376**, 399
Chaboyer, B., Demarque, P., Kernan, P.J. and Krauss, L.M. 1996, *Science*, **271**, 957
Dunlop, J. *et al.* 1996, *Nature*, **381**, 581
Feast, M.W. and Catchpole, R.W. 1997, *M.N.R.A.S.*, **286**, L1
Hamann, F. and Ferland, G. 1993, *Ap.J.*, **418**, 11
Hamann, F. and Ferland, G. 1992, *Ap.J.*, **391**, L53
Kawara, K., Murayama, T., Taniguchi, Y. and Arimoto, N. 1996, *Ap.J.*, **470**, L85
Ried, I.N. 1997, *A.J.*, **114**, 161
Sneden, C. *et al.* 1996, *Ap.J.*, **467**, 819
Spinrad, H. *et al.* 1997, *Ap.J.*, **484**, 581
Tsujimoto, T., Miyamoto, M. and Yoshii, Y. 1997, *Ap.J.Lett*, in press
Tsujimoto, T. *et al.* 1995, *M.N.R.A.S.*, **277**, 945
Wills, B.J., Netzer, H. and Wills, D. 1985, *Ap.J.*, **288**, 94
Yoshii, Y., Tsujimoto, T. and Nomoto, K. 1996, *Ap.J.*, **462**, 266

STELLAR YIELDS AND CHEMICAL EVOLUTION

BRAD K. GIBSON
Mount Stromlo & Siding Spring Observatories
Weston Creek P.O., Weston, ACT, 2611 Australia

1. Introduction

Several speakers at this symposium have alluded to the zeroth-order agreement between the Type II supernovae (SNe) stellar yields, as predicted by the models of those most responsible for driving progress in the field - i.e. Arnett (1991,1996); Maeder (1992); Woosley & Weaver (1995); Langer & Henkel (1995); Thielemann et al. (1996), hereafter referred to as A91, A96, M92, WW95, LH95, and TNH96, respectively. It is important though for those entering (or indeed, already involved in!) the chemical evolution field to be cognizant of the fact that there are important first- and second-order differences between the yield compilations.

In the next few pages, I provide a qualitative comparison of the currently available Type II SNe yield grids. The strengths and weaknesses of a given grid, demonstrated by comparing against relevant observations, are noted. Some simple chemical evolution models are shown which graphically demonstrate the effect of yield grid selection.

2. Stellar Yields

The most recent oxygen and iron yield predictions, from the aforementioned modelers, are shown in Figure 1. While the general trend of oxygen ejected as a function of progenitor mass is similar – the solar metallicity M92 yields are the exception, due primarily to the inclusion of metallicity-dependent mass-loss in the evolutionary models – it is also apparent that there is a zero-point uncertainty at the level of a factor of $\sim 2 \rightarrow 3$. Because of remaining uncertainties in the treatment of convection, mass-loss, reaction rates, evolving helium cores versus self-consistent evolution from the zero age main sequence (ZAMS), etc, we simply cannot (yet!) predict oxygen yields to an accuracy better than this factor of $\sim 2 \rightarrow 3$ (Langer 1997).

K. Nomoto and J.W. Truran (eds.), Cosmic Chemical Evolution, 159–163.

While the predicted oxygen yield depends primarily upon the physics of hydrostatic burning, that of iron is more closely tied to that of explosive nucleosynthesis. Iron is more problematic than oxygen in many ways as it is linked inexorably to the placement of the mass-cut, effectively a free parameter in the stellar models. This makes the iron yields of Figure 1 *highly* uncertain; other than the "calibration" points provided by SNe 1987A and 1993J (see TNH96) at $m = 20\ M_\odot$ and $14\ M_\odot$, respectively, *a priori* predictions of iron yields at other progenitor masses should be regarded with caution.

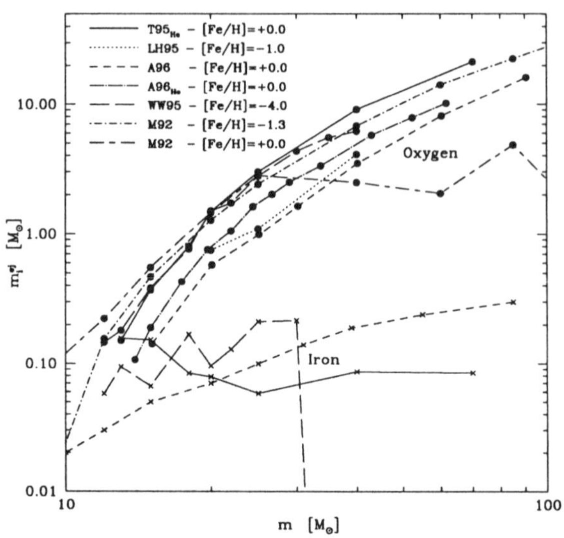

Figure 1. Oxygen and iron yields as a function of progenitor mass, for the primary Type II SNe yield compilations currently available. The models of M92, LH95, and A96 are only evolved to the completion of oxygen burning; those with a subscript 'He' appended refer to evolved helium cores only - the remainder were evolved self-consistently from the ZAMS.

Figure 2 shows the ejecta C/O ratio, as a function of progenitor mass, for a variety of model sources. Whereas oxygen by itself has a factor of ~ 3 uncertainty, the ratio of C/O is even worse. Here we can see that for $m \gtrsim 15\ M_\odot$, the uncertainty has grown to a factor of $\sim 5 \rightarrow 10$! The shaded region represents the Reimers' et al (1992) claim of log C/O≈ -1 in high redshift Lyman-limit systems. It should be readily apparent that, regardless of initial mass function (IMF) and star formation history, the A91 and LH95 yields would be hard-pressed to ever recover such a low C/O.

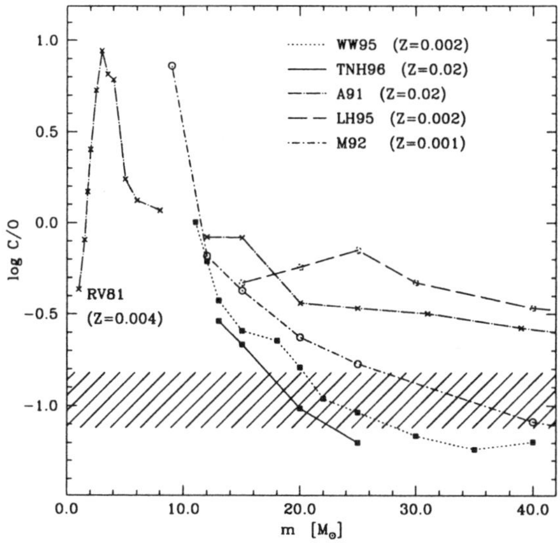

Figure 2. Ratio of carbon to oxygen in the model Type II SNe yields. Renzini & Voli's (1981) low and intermediate mass stellar yields are shown for comparison. The shaded region represents the observed C/O in high-redshift Lyman-limit systems (Reimers et al. 1992).

Figure 3 shows the predicted yield [O/Mg] and [Si/Mg], as a function of progenitor mass, overlaid on the observed Milky Way bulge giant underabundances (i.e. [O,Si/Mg]≈ -0.3), as reported by Rich (1996). Again, independent of IMF or star formation history, we can conclude that the WW95 solar metallicity yields are incompatible with Rich's (1996) bulge [O/Mg]. The TNH96 [O/Mg] are only marginally compatible, as are the WW95 [Si/Mg]. Resorting to Type Ia SNe to alleviate the "discrepancy" does not help as $[\mathrm{O,Si/Mg}]_{\mathrm{Ia}} > +0.0$.

3. Chemical Evolution

Figure 4 demonstrates how the different yield behaviors alluded to previously translate into a simple chemical evolution application. Here I show the evolution of the interstellar medium metallicity Z_g, C/O, and [O/Fe] for a massive elliptical galaxy, assuming star formation is directly proportional to the available gas mass. The star formation timescale has been adjusted to ensure the present-day photo-chemical properties of the models match local observations. All other input ingredients are identical. Despite the final (V-K) and $<Z>_*$ being the same in all cases, the behavior of individual elemental ratios can be *substantially* different!

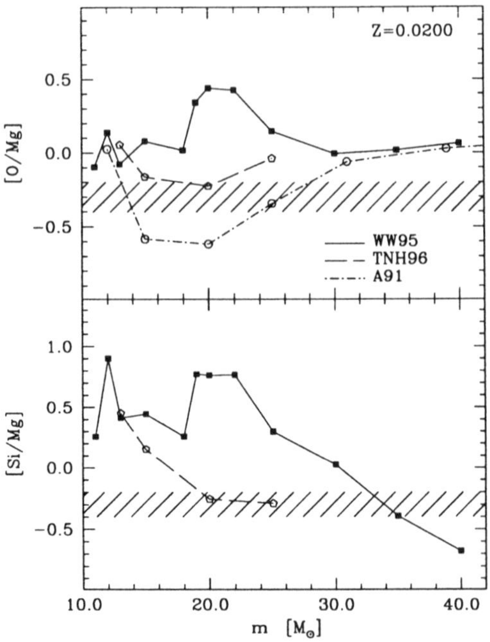

Figure 3. Comparison of the model Type II SNe yields, with the Milky Way bulge giant abundances (shaded region), as noted in Rich (1996).

4. Implications

My only goal in this contribution has been to present a cursory comparison of a number of Type II SNe yield contributions, demonstrating through simple non-quantitative figures that there still exist significant differences in the predictions available in the literature. One yield package which is entirely suited to one particular chemical evolution application may be entirely incompatible with another! Thomas et al. (1997) have recently come to the same conclusion.

Finally, we wish to note here that the aforementioned uncertainties in the massive star stellar yields make determining an (accurate) upper mass limit to the IMF (Gibson 1998) and an (accurate) accounting of the Type II SNe ICM iron fractionary contribution (Gibson et al. 1997) a difficult proposition, despite recent claims to the contrary.

5. References

Arnett, D. 1991, Frontiers of Stellar Evolution, ed. D.L. Lambert, ASP Conf. Series, 389 (A91)

Arnett, D. 1996, Supernovae and Nucleosynthesis, Princeton: Princeton

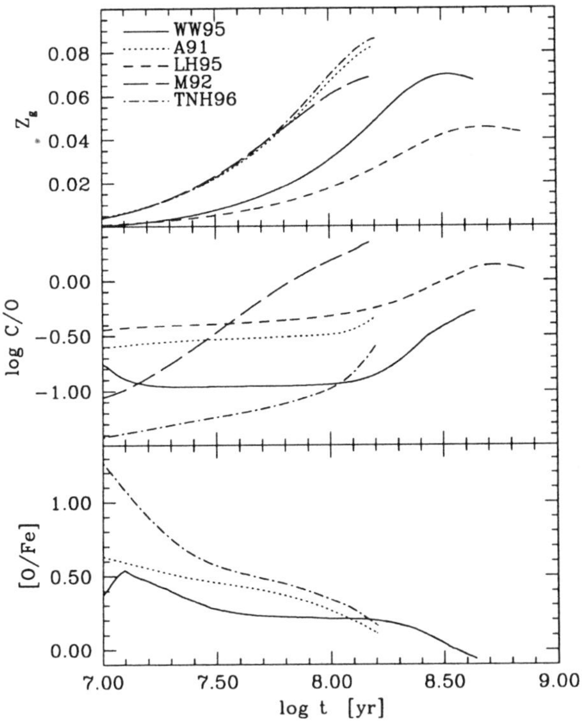

Figure 4. Evolution of the ISM metallicity Z_g, C/O, and O/Fe ratios as a function of time. The timescale for star formation was varied in order to recover a (V-K) and luminosity-weighted mean metallicity in agreement with that observed locally. Star formation is assumed to cease with the onset of a galactic wind.

Univ. Press (A96)

Gibson, B.K. 1998, ApJ, submitted

Gibson, B.K., Loewenstein, M. & Mushotzky, R.F. 1997, MNRAS, 290, 623

Langer, N. 1997, The History of the Milky Way and its Satellite System, ed. A. Burkert, et al. , ASP Conf. Series, 169

Langer, N. & Henkel, C. 1995, Space Sci. Rev., 74, 343 (LH95)

Maeder, A. 1992, A&A, 264, 105 (M92)

Reimers, D., et al. 1992, Nature, 360, 561

Renzini, A. & Voli, M. 1981, A&A, 94, 175 (RV81)

Rich, R.M. 1996, New Light on Galaxy Evolution, ed. R. Bender & R.L. Davies, Dordrecht: Kluwer, 19

Thielemann, F.-K., Nomoto, K. & Hashimoto, M. 1996, ApJ, 460, 108 (TNH96)

Thomas, D., Greggio, L. & Bender, R. 1997, MNRAS, in press

Woosley, S.E. & Weaver, T.A. 1995, ApJS, 101, 181 (WW95)

THE FORMATION OF FIRST GENERATION STARS AND GLOBULAR CLUSTERS IN PROTOGALACTIC CLOUDS

D.N.C. LIN

Lick Observatory, Univ. of California, Santa Cruz, CA 95064

AND

S.D. MURRAY

Lawrence Livermore Nat'l Lab., L-23, Livermore, CA 94550

Within collapsing protogalaxies, thermal instability leads to the formation of a population of cool fragments which are confined by the pressure of a residual hot background medium. The critical mass required for the cold clouds to become gravitationally unstable and to form stars is determined by both their internal temperature and external pressure. Massive first generation stars form in primordial clouds with sufficient column density to shield themselves from external UV photons emitted by nearby massive stars or AGNs. Less massive photoionized clouds gain mass as they undergo cohesive collisions with each other, and lose mass due to ram pressure stripping by the residual halo gas. Collisions may also trigger thermal instability and fragmentation into cloudlets. While most cloudlets have substellar masses, the largest become self-gravitating and collapse to form protostellar cores without further fragmentation. The initial stellar mass function is established as these cores capture additional residual cloudlets. Energy dissipation from the mergers ensures that the cluster remains bound in the limit of low star formation efficiency. Dissipation also promotes the formation and retention of the most massive stars in the cluster center. On the scale of the protogalactic clouds, the formation of massive stars generates intense UV radiation which photoionizes gas and quenches star formation in nearby regions. As gas density accumulates in the center of the the galactic potential, the self regulated star formation rate increases. At the location where most of the residual gas can be converted into stars on its internal dynamical timescale, a galaxy attains its asymptotic kinematic structure such as exponential profiles, Tully-Fisher, and Faber-Jackson laws.

K. Nomoto and J.W. Truran (eds.), Cosmic Chemical Evolution, 165–174.

1. Fragmentation of Protogalactic Clouds

During the collapse of a protogalactic cloud (PGC), density inhomogeneities and velocity variations lead to shocks which heat the gas to the virial temperature ($\sim 10^{6-7}$ K) of the galactic halo (Binney 1977; Rees & Ostriker 1977; White & Rees 1978). In order for a PGC to collapse, its cooling time scale, τ_c, must be shorter than the dynamical time scale, τ_d, on which it can contract. For PGC's with masses comparable to the Galaxy, this condition is satisfied when their length scale $D < 100$ kpc (Blumenthal *et al.* 1984).

The emergence of small scale structure requires that either the growth rate or the limiting amplitude of any perturbations be decreasing functions of their length scale. In a non rotating, cold, homogeneous cloud, gravitational instability alone cannot induce fragmentation, because its growth rate is essentially independent of length scale. In the limit that $\tau_c < \tau_d$, thermal instability can lead to the rapid growth of perturbations from infinitesimal to nonlinear amplitudes (Field 1965). At the virial temperature of the PGC, the dominant cooling mechanisms are bremsstrahlung and recombination processes (Dalgarno & McCray 1972) for which τ_c increases with temperature. In this case, the temperature difference between cooler perturbed regions and the background is amplified leading to the formation of a two phase medium.

Although thermal instability proceeds more rapidly than the collapse of the cloud, its growth rate is essentially independent of length scale, because it is determined by the local cooling rate. Nonetheless, the growth of over dense regions in a thermally unstable cloud is limited by the transition from isobaric to isochoric evolution, and the limiting amplitude is a decreasing function of the length scale (Burkert & Lin in preparation). Across the interface between the perturbed region and the background, differential cooling leads to a pressure gradient which induces gas flow from the hot background towards the cooler perturbed regions. The density enhancement in the cooler region reduces τ_c/τ_d in the cooler gas, and reduces τ_c in the cooler gas further below that in the background. The pressure-driven acceleration of the interface also induces the onset of Rayleigh-Taylor instability. The growth rate of Rayleigh-Taylor instability increases for small wavelengths, λ. The growth rate of the small λ perturbations, however, decrease when they become nonlinear, ie. when their amplitudes $a \approx \lambda$, and later evolution is dominated by perturbations with larger wavelengths.

The above discussions indicate that although thermal instability can lead to runaway growth of perturbations, it does not lead to their breakup to small length scales. We now examine the processes which can lead to the fragmentation of the cold regions.

Once the final T is attained, pressure balance is re-established on all scales and the PGC becomes a two-phase medium with density (n) contrast inversely proportional to the temperature (T) ratio. The residual halo gas (RHG) in the background remains at the virial temperature, with a density such that its thermal energy is lost through both radiative cooling and thermal conduction between it and the cold clouds (McKee & Cowie 1977). Due to buoyancy effects, the clouds fall toward the center of the galactic potential and dissipate its gravitational to heat the RHG. At a galactocentric radius $D \sim 10$ kpc, an energy balance is attained with $nT \sim 10^{3-4}$ (Lin & Murray in preparation).

The cooling efficiencies of bremsstrahlung, recombination, and atomic hydrogen emission decrease rapidly below $\sim 10^4$ K. In metal-free gas, however, non-equilibrium recombination leads to the formation of a small amount of H^- ions which combine with neutral H to form H_2. Emission by H_2 is capable of reducing the gas temperature, T to $\sim 10^2$ K (Murray & Lin 1990). In metal-enriched clouds, emission by metal ions, such as CII, molecules such as CO, and grains provide additional cooling to lower equilibrium temperatures. Once the final temperature is attained, pressure balance is re-established on all scales.

The clouds' infall motion is affected by both their coagulation and RHG's drag (Lin & Yorke in preparation). The motion of the clouds through the RHG also leads to mass loss due to Kelvin-Helmholtz instability (Murray *et al.* 1993). Since the growth time scale of Kelvin Helmholtz instability increases with the perturbation wavelength, it leads to fragmentation. The break down of the clouds increases their collective area filling factor and so increases their collision frequency. A balance between disruption and coagulation establishes an equilibrium size distribution.

2. Formation of Protocluster Clouds

A lower limit on their size distribution is set by the clouds' evaporation due to conduction from the hot RHG. At higher masses, the central density of the clouds increases with mass. Above a critical mass, M_c, thermal pressure can no longer support the weight of the envelope (Bonner 1956), and the clouds undergo inside-out collapse (Shu 1977). The inside-out collapse is stable and does not lead to fragmentation without any further unstable cooling. Contrary to the opacity-limited fragmentation scenario (Hoyle 1953; Low & Lynden-Bell 1976), M_c represents the minimum mass for isothermal collapsing clouds (Tsai, in preparation).

In primordial gas, H_2 emission leads to $T \sim 10^2$ K, which corresponds to $M_c \sim 10^2 M_\odot$, and the formation of massive stars. A population of $\sim 10^4$ O5 stars produce adequate UV radiation to photoionize the entire PGC out to 100 kpc, which leads to $T \sim 10^4$ K and $M_c \sim 10^6 M_\odot$. Since M_c is comparable to the mass of globular clusters, we refer these warm, massive clouds as protocluster clouds (PCC's). Clouds with $M < M_c$ are stable, and star formation is quenched. As some of the massive stars evolve off the main sequence, the UV flux diminishes, cooling again leads to $T \sim 10^2$ K in sheltered regions, and spontaneous star formation resumes. This self-regulated star formation scenario has three implications: 1) Stars formed in a metal-poor environment are massive and short-lived, consistent with their rarity today. 2) The elemental abundance distribution is produced by type II supernovae, consistent with that observed among stars with [Fe/H]< -1 (Wheeler *et al.* 1989). 3) The self-regulated star formation rate naturally yields [Fe/H] ~ 0.1 on the collapse time scale of the PGC, $\tau_d \sim 1$ Gyr, consistent with the halo stars.

We can estimate nT of the PCC's from the values of n and the velocity dispersion (σ) of Galactic globular clusters, averaged over their half-mass radius (r_h). These quantities do not change during the post formation evolution. The clouds' radius prior to star formation is probably $> 2r_h$ but smaller than their tidal radii (several r_h), such that their initial density may be 1-3 orders of magnitude smaller than the average cluster density at r_h today. From σ we estimate $T \sim 10^4$ K, comparable to that expected if they were photoionized. Since both the magnitude ($\sim 10^{2-5}$), and the dependence on the Galactic distance ($\propto D^{-3}$) inferred for nT are consistent with those expected for the RHG (Murray & Lin 1992), we postulate that PCC's with mass $< M_c$ are confined by the RHG's pressure.

From these results, and the cluster metallicities, we can also estimate the cooling time scale (τ_{cc}) and dynamical time scale (τ_{dc}), of the PCC's. The ratio τ_{cc}/τ_{cd} increases from $\sim 10^{-4}$ near the Galactic bulge to ~ 1 at ~ 100 kpc. In most PCC's $\tau_{cc} << \tau_{cd}$ and thermal equilibrium is only possible in the presence of external UV photons with a flux comparable to that of a single O5 star at a distance comparable to the PCC's size (typically a few pc). This flux is also comparable to that required for self regulated star formation in the halo.

3. Induced Star Formation in PCC's

A constraint on the duration of the star formation epoch of a PCC can be imposed from the metal homogeneity (primarily in [Fe/H]) among stars within individual clusters. In M92, [Fe/H]$= -2.2$ and Δ [Fe/H]< 0.1 among the cluster stars (Stetson & Harris 1988). The total metal content in M92 ($\sim 30 M_\odot$) can be produced by 2 or 3 supernovae. If star formation in M92 occurred over a timescale of more than a few 10^6 years, supernovae would increase both [Fe/H] and Δ [Fe/H] to values much larger than observed.

The small Δ[Fe/H] implies that PCC's are enriched and well-mixed prior to a rapid burst of star formation. The most efficient mixing process within the PCC's is large scale turbulence (Murray & Lin 1990), which may be excited by Kelvin-Helmholtz instability at the interface with the RHG. Since shock dissipation damps supersonic motion efficiently, the mixing time scale is typically a few $\tau_{dc} \gg \tau_{cc}$. External heating by nearby massive stars is then required to maintain a stable thermal equilibrium for the PCC's. Such an equilibrium is only stable, however, if the PCC's are not strongly self-gravitating.

In PCC's with marginal self-gravity, mass loss due to ram pressure stripping by RHG is stabilized. Mass loss may also be compensated by coagulation in a collisional equilibrium. Collisions also introduce perturbations which, with a sufficiently large amplitude, could increase the cooling efficiency, leading to star formation (Murray & Lin 1989). In isolated, but relatively massive PCC's, where self-gravity is important, external heating can no longer stabilize thermal instability (Murray & Lin 1992), and runaway cooling leads to $T <$100 K. The corresponding n enhancement would increase the recombination rate and opacity. With sufficiently large column density, self-shielding may also act to prevent the heating of the PCC's interior. All of these effects indicate that there is a small range of mass over which the PCC's become thermally unstable and begin to collapse.

While most clusters are members of the spheroidal population, the average value of D is smaller than that of the Sun. Dissipative collisions between PCC's reduce their specific energy to values comparable to disk stars, much below that of the PGC. Collisions between two PCC's also lead to the separation of gas from dark matter and older stars which may have formed within them and contributed to their prior enrichment.

The onset of thermal instability, and the subsequent Rayleigh-Taylor and Kelvin-Helmholtz instabilities caused from the resulting collapse of the PCC's induce them to fragment into a population of small cloudlets analogous to the formation of the PCC's within a PGC. The small cloudlets are pressure confined by the warm (10^4K) residual gas (WRG) which is, in turn, in pressure balance with the RHG. Due to their collective gravity, and their buoyancy in the WRG, the cloudlets collapses towards the center of PCC. During their infall, the cloudlets lose mass through ram pressure stripping and gain mass through coagulation. Cloudlets with masses $> M_c$ collapse to form protostellar cores. Since no additional fragmentation is expected, the minimum stellar mass is $\sim M_c \propto T^2$.

We cited above circumstantial evidence that the timescale for star formation in clusters is less than a few million years. On such a timescale, cloudlets with adequate metallicity ([Fe/H]> -2.5) can cool to 10 K and form low mass stars before the emergence of massive stars which would emit intense UV radiation, photoionize cloudlets, and quench star formation.

Since their cooling timescale is > 10^6 yr, extremely metal poor cloudlets ([Fe/H]< −2.5) attain a minimum temperature > 10 K before being photoionized. Thus, metal poor stars tend to be relatively massive and short lived accounting for their rarity today (Hellsten & Lin 1998).

Numerical simulations of the coagulation of cloudlets in PCC's (Murray & Lin 1996) show that initially the geometric cross section determines the collisional frequency between uncollapsed cloudlets, leading to a relatively flat mass distribution. After the cloudlets acquire sufficient mass to form protostellar cores, they continue to acquire additional mass as they merge with residual cloudlets. The increasing mass and compact sizes of the cores, causes their capture rate of residual cloudlets to be strongly affected by gravitational focusing. As a result, their growth timescale decreases with mass. A few massive cores rapidly grow prior to any significant mass increase among most other cores, leading to a steep initial mass function (IMF) for the cores. The cloudlet-core collisions also lead to the dissipation of their relative motion, while global instabilities increase their velocity dispersion (Aarseth *et al.* 1988). In the final stage of the PCC's collapse, the velocity dispersion increases rapidly, and the collisional frequency is once again determined by the geometrical cross section of the cloudlets, leading to a flattening of the IMF during the collapse. The final slope of the IMF is determined by whether or not most cloudlets have undergone collisions to form protostellar cores before gravitational focusing ceases to be important.

The growth of the most massive cores is terminated when their UV emission heats and ionizes nearby cloudlets. The most massive stars require many dissipative mergers, and so are preferentially formed and tend to remain in the cluster center. This expectation is in contrast to the implication of opacity-limited fragmentation scenario, in which the Jean's criterion suggests that the high mass stars are preferentially formed in low density environments.

After the first generation of stars are formed, the residual gas must be removed from the clusters to avoid self-contamination. Extrapolating from the present stellar population, a Salpeter IMF would imply the existence of tens of massive stars within a typical globular cluster. The UV flux from these massive stars is sufficiently intense to photoionize the residual gas throughout a PCC (Tenorio-Tagle *et al* 1986). In the shallow potential of PCC's, the resultant internal heating leads to the formation of an expanding ionization wave which efficiently clears out the residual gas within a few 10^6 yr (Noriega-Crespo *et al.* 1989).

Unless more than half of the gas in PCC's is converted into stars, the removal of the gas on a timescale comparable to the dynamical timescale rapidly reduces the depth of the cluster's potential. In most cases, the disposal of residual gas would then lead to the disruption of the cluster (Lada *al.* 1984). If, however, young stars form through the coagulation of low mass cloudlets, energy dissipation resulting from the mergers leads to the resulting star clusters having radii much smaller than the original PCC's. It is therefore more likely that the newly formed clusters will remain gravitationally bound even in the limit of inefficient star formation. Numerical simulations show up to an order of magnitude reduction in the half mass radius of the star cluster formed through coagulation of a system of collapsing cloudlets (Murray & Lin 1996).

4. Self-Regulated Star Formation in Collapsing Protogalactic Clouds

Various models of star formation during the collapse of a PGC have been constructed in an attempt to match the galactic stellar distributions and chemical enrichment history with observational data (Larson 1974). The results of these investigations show that the observations are best simulated with a star formation law in which the star formation rate $\propto n^2$, This empirical prescription is essentially *ad hoc* and its direct comparison with the observations of current start formation rates in the Galactic disk (Schmidt 1956) may be inappropriate, since the physical conditions in PGCs were probably very different from those present day.

Based on the above considerations, the collapse of a PGC may, in principle, be accompanied by an epoch of efficient star formation in which a large fraction of the PGC is turned into stars. The outcome of such an event would be the formation of galaxies with radii which are much larger than typically observed Holmberg radii. One possible mechanism for suppressing star formation is through external heating due to background UV radiation sufficient to photodissociate H_2 molecules, as discussed above in the context of globular clusters.

At large galactic distance ($D > 30$kpc), $n < 10^{-1}$ for clouds with masses $< M_c$ and $T > 10^4$ because $nT < 10^{2-3}$ in the RHG. The cooling rate cannot exceed the rate of photoionization and dissociation resulting from the UV flux of external AGNs. Inside 30 kpc, however, adequate UV radiation can only be provided by massive stars formed within the PGC (Lin & Murray 1992). When the population of massive stars reaches a critical number, they can photoionize and dissociate the entire PGC. Subsequent star formation is suppressed until the massive stars evolve off the main sequence. Both the internal density and the filling factor of the clouds are decreasing functions of D. Thus, the recombination rate, the critical UV flux, star formation rate, and heavy element yield all decrease at larger D. At some small radii, even with all the residual gas converted into stars on a dynamical timescale, there is insufficient number of massive stars to provide sufficient UV radiation to quench star formation. At that point, efficient star formation ensues. The dissipational gaseous collapse at larger D becomes, instead, dissipationless stellar collapse, and the protogalaxy become a bona fide galaxy. The continuous formation of stars leads to a luminosity $\sim 10^{43-44}$ ergs s^{-1} and the gas is enriched to [Fe/H]> -2. Dissipation during the collapse implies that stars will retain a memory of the point at which they formed, leading to the formation of steep metallicity gradients. After virialization, the final radii are comparable to the typical scale lengths observed in galaxies. In the absence of rotation, the mass dependence of the critical radius leads naturally to the Faber-Jackson relation of elliptical galaxies (Lin & Murray 1992).

Similarly, in disk galaxies, rapid star formation is prevented unless the surface density exceeds a critical value. This condition cannot be reached beyond a radius at which the disk becomes gravitationally unstable before the onset of rapid star formation. This transition radius is comparable to the scale length of exponential disks and its dependence on the galactic mass is similar to that of the Tully-Fisher law. At larger D, star formation is self regulated and inefficient, resulting in the observed exponential profiles (Lin & Pringle 1987).

5. Comparison With Present-Day Cluster and Star Formation

In the construction of the above comprehensive model for the formation of first generation stars and stellar clusters, we used the observed dynamical and chemical properties of globular clusters as constraints. This model has many implications on the observational properties of young clusters. However, young globular clusters are only found in distant galaxies such as NGC 1275 (Holtzman *et al.* 1992), where they cannot presently be resolved. However, recent observations find that most stars in the Galaxy today form in clusters in which the timescale for star formation is quite short (Lada et al. 1991; Lada 1992). The central density of some young clusters are comparable to that of some globular clusters. In the Trapezium cluster, all the stars appear to have ages $< 10^6$ yr (Prosser *et al.* 1994) which is $\sim 2\tau_{dc}$. These timescales are consistent with the above model, in which star formation proceeds through a sequence of initial gas fragmentation, coagulation, protostellar collapse, and the clearing of residual gas. (There are exceptions such as IC 348, where star formation has persisted for many τ_{dc}, Lada & Lada, 1995). It is therefore of value to compare the implications of our model with the dynamical properties of nearby sites of ongoing star formation, to provide both supporting evidence and constraints upon the above picture of star formation.

Molecular clouds are clumpy on all scales (Scalo 1985; Zinnecker *et al.* 1993). Complex cloud substructure may also be inferred from the large dispersion, over a small field, in the observed extinction of the stars in the background cluster IC 5146 (Lada *et al.* 1994). (The extinction is equivalent to the surface density of the intervening clouds). These observations suggest that fragmentation occurs in the clouds prior to the gravitational collapse of individual protostellar cores as we have postulated above.

Magnetic fields, neglected in our cluster and star formation scenario, are observed to be important in regulating the structure of star forming regions today. In these regions, the velocity dispersions, which are correlated with the length scale of the substructures (Larson 1981), are often larger than the sound speed inferred from the transition temperature, but are comparable to the Alfvén speed (Heiles *et al.* 1993; Caselli & Myers 1995). In some regions, the dispersive motion of the clouds may be regulated by the interstellar magnetic fields. There are also, however, magnetic supercritical regions, where the magnetic field can no longer balance gravity. These are the regions where massive stars and small clusters are formed. The lack of polarization in the densest cores of molecular clouds suggest that magnetic fields are excluded from these regions (Goodman *et al.* 1995). The decoupling of the field from the protostellar clouds is equivalent to the loss of thermal support during a cooling instability, and may also lead to complex substructures (Terquem & Lin in preparation).

The mass functions of dark cloudlets in star forming complexes such as Ophiuchus, Taurus, Orion (Scalo 1985), and L1630 (Lada *et al.* 1991) have very similar power-law distributions, and are considerably flatter than that of the stellar IMF. The extrapolated collisional timescale for the small cloudlets is comparable to a few local dynamical timescales, and so their size distribution could arise naturally from a collisional equilibrium. The relatively flat mass spectrum is then consistent with that obtained from numerical simulations of the coagulation processes among protostellar cores (Murray & Lin 1996), and implies that the physical cross section determines the merger rate among the cloudlets. In contrast, gravitational focusing is more important for the capture of residual cloudlets by collapsed stellar cores, leading to a steeper stellar IMF.

Shortly after the formation of protostellar cores, they are surrounded by the residual gas and appear as embedded sources. IR observations indicate that the embedded sources are strongly clustered (Greene *et. al.* 1994). These clusters are much more centrally condensed than the host cloud complex. Furthermore, the brightest embedded sources are usually found at the center of the clusters (Lada 1992). The luminosity segregation is consistent with the concept that protostellar cores form through dissipative mergers of small cloudlets, such that the most massive stars preferentially form in regions of high density where the collision frequency is greatest.

Finally, in older star forming regions, the young stellar objects emerge as T Tauri stars. In the Orion Nebula, T Tauri stars also appear to be clustered. The higher luminosity T Tauri stars are more centrally condensed than those with low luminosities (Prosser *et al.* 1994). In these regions, there is not sufficient time for post-formation dynamical evolution toward mass segregation. The observations are consistent with the more massive stars forming in dense, central regions. Since most of their kinetic energy is dissipated during the coagulation, these massive stars remain near PCC's center after the residual gas is cleared.

References

Aarseth, S.J., Lin, D.N.C., & Papaloizou, J.C.B. 1988, ApJ, 324, 288
Binney, J. J. 1977, ApJ, 215, 483
Blumenthal, G. R., Faber, S. M., Primack, J. R., & Rees, M. J. 1984, *Nature*, 311, 517
Bonner, W. B. 1956, MNRAS, 116, 356
Caselli, P. & Myers, P.C. 1995, ApJ, 446, 665
Dalgarno, A., & McCray, R. A. 1972, ARAA, 10, 375
Field, G. B. 1965, ApJ, 142, 531
Goodman, A.A., Jones, T.J., Lada, E.A., Myers, P.C. 1995, ApJ 448, 748
Greene, T.P., Wilking, B.A., André, P., Young, E., & Lada, C.J. 1994, ApJ, 434, 614
Heiles, C., Goodman, A.A., & McKee, C.F. 1993 in *Protostars and planet III*, eds. E. H. Levy & J. I. Lunine (Tucson: Univ. Arizona Press), 279
Holtzman, J. A., *et al.* 1992, AJ, 103, 691
Hoyle, F. 1953, ApJ, 118, 513
Lada, C. J., Lada, E. A., Clemens, D. P., & Bally, J. 1994, ApJ, 429, 694
Lada, C. J., Margulis, M., & Dearborn, D. 1984, ApJ, 285, 141
Lada, E. A. 1992, ApJL, 1992, 393, L25
Lada, E. A., DePoy, D. L., Evans, N. J., & Gatley, I. 1991, ApJ, 371, 171
Lada, E.A. & Lada, C.J. 1995, AJ, 109, 1684
Larson, R. B. 1974, MNRAS, 169, 229
— 1981, MNRAS, 194, 809
Lin, D. N. C., & Murray, S. D. 1992, ApJ, 394, 523
Lin, D. N. C., & Pringle, J. E. 1987, ApJL, 320, L87
Low, C., & Lynden-Bell, D. 1976, MNRAS, 176, 367
McKee, C. F., & Cowie, L. L. 1977, ApJ, 215, 213
Murray, S. D., & Lin, D. N. C. 1989, ApJ, 339, 933
— 1990, ApJ, 357, 105
— 1992, ApJ, 400, 265
— 1996, ApJ, 467, 728
Murray, S. D., White, S. D. M., Blondin, J. M., & Lin, D. N. C. 1993, ApJ, 407, 588
Noriega-Crespo, A. Bodenheimer, P. Lin, D. Tenorio-Tagle, G. 1989, MNRAS, 237, 461
Prosser, C. F. *et al.* 1994, ApJ, 421, 517
Rees, M. J., & Ostriker, J. P. 1977, MNRAS, 179, 541
Scalo, J.M. 1985, in *Protostars and Planets II*, eds. D.C. Black & M. Matthews, (Univ. of Arizona Press), 201
Shu, F. 1977, ApJ, 214, 488
Stetson, P. B. & Harris, W. E. 1988, AJ, 96, 909
Tenorio-Tagle, G. Bodenheimer, P. Lin, D. Noriega-Crespo, A. 1986, MNRAS, 221, 635
Wheeler, J. C., Sneden, C., & Truran, J. W. 1989, ARAA, 27, 279
White, S. D. M., & Rees, M. J. 1978, MNRAS, 183, 341
Zinnecker, H., McCaughrean, M., & Wilking, B. A. 1993, in *Protostars and Planets III*, eds. E. Levy & J. Lunine, (Univ. of Arizona Press), 429

CONSTRAINTS ON GALAXY FORMATION FROM EXTRAGALACTIC GLOBULAR CLUSTERS

JEAN P. BRODIE
Lick Observatory
University of California, Santa Cruz, CA 95060, U.S.A.

1. Introduction

The merger model for elliptical galaxy formation has received increasing attention since it was first suggested by Toomre & Toomre (1972). Van den Bergh (1984) pointed out a problem with the idea that elliptical galaxies were formed by simply combining two, or more, spiral galaxies. He noted that the specific frequency (S_N, number of globular clusters per unit galaxy light) is systematically lower for spirals than for ellipticals. Schweizer (1987) suggested that globular clusters might be expected to form in the merger process, thereby alleviating or possibly eliminating the S_N problem. Ashman & Zepf (1992) developed this idea into a merger model for globular cluster formation with testable predictions.

We recently examined this model in the light of new HST and ground–based imaging data on the blue and red sub–populations of globular clusters in elliptical galaxies (Forbes, Brodie & Grillmair 1997). We concluded that the merger model for globular cluster formation has serious problems, especially in explaining the characteristics of globular clusters in giant elliptical galaxies with high S_N. For example, the globular cluster systems of high S_N galaxies do not have more metal–rich globular clusters as would be expected in the merger model but instead have more metal poor ones. This means that globular cluster metallicity gradients are steeper in high S_N galaxies, contrary to the predictions of Ashman & Zepf (1992).

We proposed a multi–phase collapse scenario, more consistent with the available evidence (see also Kissler-Patig 1997). We speculated that there are three main phases in the collapse process (which we called pre–galaxy, galaxy and disk phases) each of which may have associated globular cluster formation. The metal–poor globular clusters form at an early stage in the collapse of the protogalactic cloud, in the pre–galaxy phase. Only a small

K. Nomoto and J.W. Truran (eds.), Cosmic Chemical Evolution, 175–184.

fraction of the available gas is converted into stars, and most of these stars are located in globular clusters which formed from the chaotic merging of dense, gravitationally bound clumps. In the second phase, after the gas cloud has undergone further collapse, field star formation is preferred over star formation in globular clusters (possibly due to more efficient cooling at higher metallicities). Both field stars and globular clusters will have essentially the same metallicity (i.e. relatively metal–rich). Thus the metal–rich globular clusters formed out of more enriched gas, roughly contemporaneously with the galaxy stars. The disk globular clusters in spirals may represent a third phase of this formation process. Tidal stripping of the globular clusters from nearby galaxies may have contributed to the globular cluster population in the outer parts of cD galaxies, and recent mergers may have contributed to the globular cluster population in the centers of elliptical galaxies, but we suggest that the vast majority of globular clusters in gE and cD galaxies have probably formed *in situ.*

Below we describe two "case studies" which illustrate how optical spectroscopy with the new generation of 10m–class telescopes can help constrain both globular cluster and galaxy formation models. Some of the issues that can be addressed with spectroscopic data include the following:

• Do globular clusters in other galaxies show the same abundance ratios as those in the Local Group?

• Are globular clusters in elliptical galaxies more metal rich than in spirals, as would seem to be indicated by their colors?

• What is the upper limit on [Fe/H]?

• Can abundances, abundance ratios, ages and kinematics identify globular clusters that formed during a merger?

Optical broad–band colors suffer from a well–known age–metallicity degeneracy and so have been of limited value in addressing these issues.

The first case study is an examination of the globular clusters in NGC 1399. The second is a spectroscopic study of the proto–globular cluster candidates in NGC 1275.

2. The NGC 1399 Globular Cluster System

NGC 1399 is the cD galaxy at the center of the Fornax cluster at a distance of 19.3 Mpc (Madore, et al. 1997). Several studies (Ostrov et al. 1993, Kissler-Patig et al. 1997, Forbes et al. 1997) have demonstrated the existence of bimodal, and possibly trimodal, globular cluster color distributions in this galaxy. Using the Keck telescope, we recently obtained 5 Å resolution spectra of 18 globular clusters in NGC 1399 (Kissler-Patig et al. 1998). Figure 1 shows three representative spectra of objects in the blue, intermediate and red sub–populations.

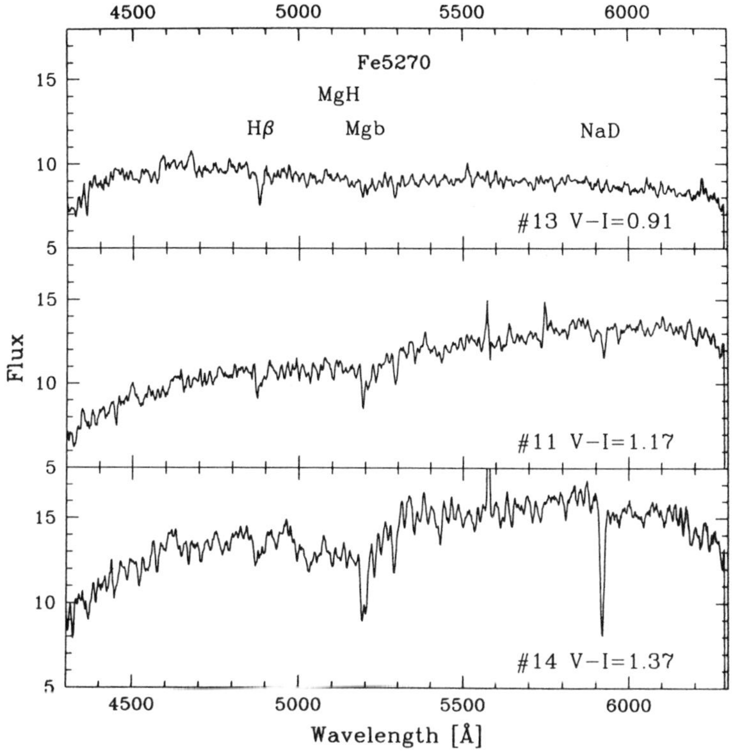

Figure 1. **Fig. 1. – Three representative spectra of globular clusters from the blue,** intermediate and red sub–populations **in NGC 1399. The spectra were smoothed over** 3 pixels. While H_β gets slightly weaker from the blue to the red object, the metal lines (Mg, Fe) become much stronger.

In general, we found no anomalies in the strengths of absorption line indices sensitive to metals (e.g. Mg, Fe), nor in the more age–sensitive Balmer–line indices (Figure 2). However, our sample contained two clusters for which: 1) Mg, Fe and Na were stronger than in any Milky Way or M31 globular clusters. These clusters may possibly also have a Mg/Fe over–abundance, similar to that seen in giant elliptical galaxies. Worthey et al. (1992) discussed possible explanations for the Mg/Fe enhancement in galaxies, perhaps the most plausible of which was a variable (flatter) IMF. 2) Hβ and Hγ (effectively the only two Balmer lines in the spectral range) were abnormally strong. A similar effect was also seen in a few M31 globular clusters (Brodie & Huchra 1991), although at a lower metallicity. The Balmer lines are too strong to be reproduced by the spectral synthesis models (Worthey 1994, Fritze v. Alvensleben & Burkert 1995) at any age or metallicity. A component of blue horizontal branch stars might offer at least a partial explanation. Such stars are not represented in the models

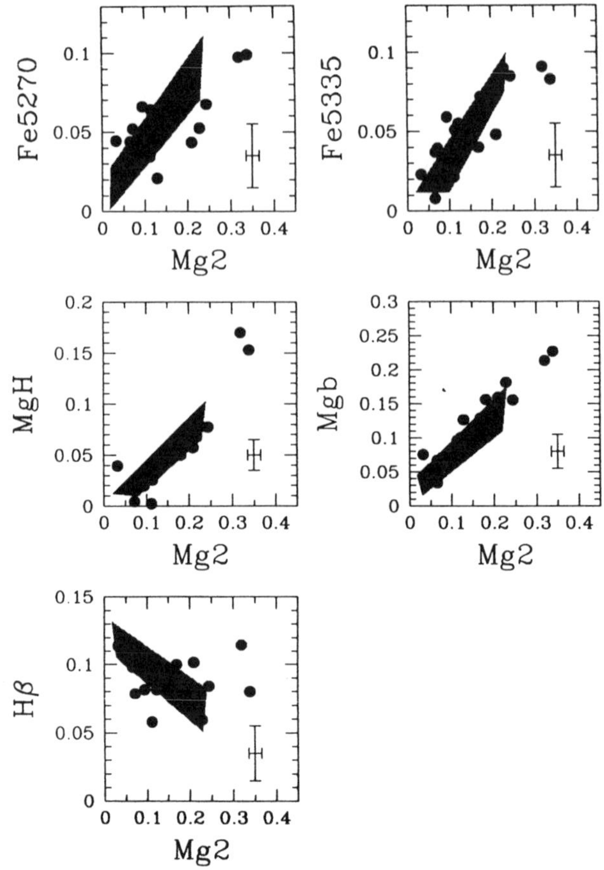

Figure 2. **Fig. 2. – Mg2 versus various other indices, with the range spanned by Milky** Way and M31 globular clusters shown as shaded regions (taken from Brodie & Huchra 1990 and Burstein et al. 1984). Apart from two very metal–rich objects, all globular clusters are consistent with abundances found in the Milky Way and M31 globular clusters.

because they were not thought to be present in metal–rich clusters. However, Rich et al. (1997) recently discovered blue horizontal branches in two metal–rich globular clusters in the Milky Way, suggesting that a revision of the models may be in order.

Since the two "anomalous" NGC 1399 clusters are also very metal rich (see below), they must have formed from solar metallicity (i.e. enriched) gas, which may indicate that they are also younger than the rest. Judging by the photometric color distribution (these peculiar clusters are in the extreme red tail of the distribution), these clusters constitute less than 5% of the total globular cluster population.

We calculated mean metallicity, [Fe/H], for all 18 clusters using the method described in Brodie & Huchra (1990). The Mg line-strengths were first corrected for a non–linear dependence of Mg index strength on metallicity for metallicities near and above solar, according to the models of Worthey (1994). Except for the two peculiar clusters, the range in metallicity seen in the NGC 1399 globulars is very similar to the range seen in the Milky Way and M31: The metal–poor clusters have metallicities similar to those of the Milky Way halo and the metal–rich clusters have metallicities similar to Milky Way bulge/disk globular clusters. Although we are unable to estimate age to better than a factor of two from our data, it appears that the vast majority of NGC 1399 globular clusters formed from processes similar to those which formed the Milky Way and M31 globular cluster systems, and at roughly the same epoch (i.e. they are old).

V–I was found to correlate well with [Fe/H] and, although individual metallicities cannot be determined to better than 0.5 dex, a mean metallicity for a globular cluster ensemble can be determined to about 0.3 dex from broad–band colors. However, the slope of the V–I vs. metallicity relation for the combination of Milky Way and NGC 1399 globular clusters was found to be twice as flat as the relation derived from Milky Way globular clusters alone (Figure 3). The consequence of this result is a much more homogeneous picture for extragalactic globular cluster systems. Metallicity estimates for very red globular clusters are now roughly solar, rather than very super solar as was previously thought, and metallicity differences between galaxies' globular cluster systems are much less significant.

In summary, it appears that NGC 1399 and its globular cluster system formed at early times. The same formation processes that were responsible for forming the Milky Way globular clusters probably formed the vast majority of NGC 1399 globular clusters. Only a small percentage need have formed later from solar metallicity gas, e.g. in a merger.

3. Proto–globular Cluster Candidates in NGC 1275

The discovery, with HST imaging, of proto–globular cluster candidates in NGC 1275 (Holtzman et al. 1992) was regarded by many as a major success of the merger model which predicted that newly–formed clusters should be observable in currently or recently merging systems. The NGC 1275 clusters constitute an important test of globular cluster formation models. NGC 1275 is the peculiar cD galaxy at the center of the Perseus cluster. It shows evidence for a merger history and may indeed be undergoing a merger at present. It also has one of the largest know cooling flows. Globular cluster formation in cooling flows (Fabian et al. 1984) received some support from the photometric study of the young clusters in NGC 1275 by Richer et al. (1993).

Shown in Figure 4 are spectra of 5 proto–globular cluster candidates in NGC 1275, obtained with the Keck I telescope (Brodie et al. 1997).

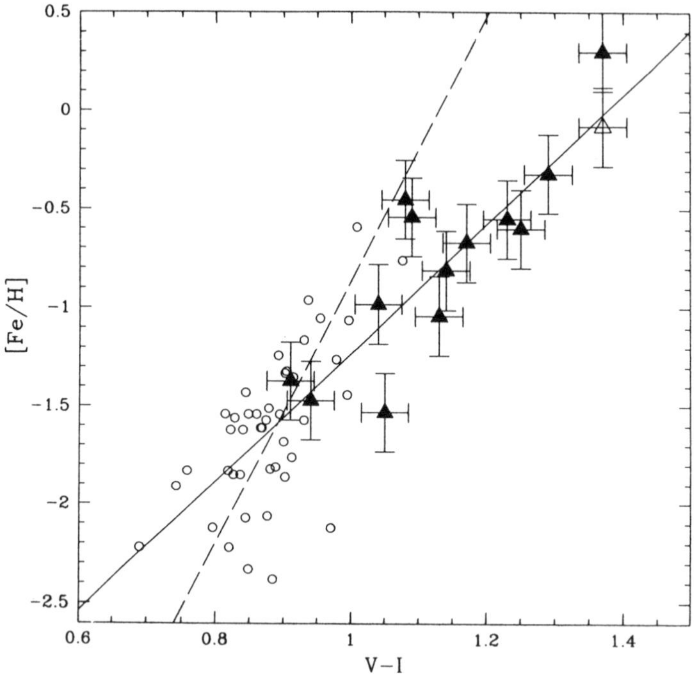

Figure 3. **Fig. 3. – V−I color versus mean metallicity of the globular clusters in NGC** 1399 (triangles). For cluster #14 the metallicity derived using only the Fe absorption line index, Fe5270, is also shown (open triangle). Over–plotted as open circles are all (de–reddened) Milky Way globular clusters with E(B−V)< 0.2 (taken from Harris 1996). The solid line is the best fit to the data, the dashed line was derived from the Milky Way sample by Kissler-Patig et al. (1997) and extrapolated to higher metallicities.

These reveal that the candidates are not HII regions, are clearly dominated by early A–type stars, and are not similar to young or intermediate age Magellanic Cloud or Milky Way open clusters.

The Balmer absorption lines were found to be too strong to be consistent with any of the standard IMF (Salpeter or Scalo), solar metallicity, Bruzual & Charlot stellar evolutionary models at any age (Figure 5). The preliminary Bruzual & Charlot (1997) models indicate that no appreciable increase in equivalent width can be achieved by changing the metallicity. However, a 2–3 $M_\odot$ IMF, adopted to simulate a flatter IMF, reproduces the observed equivalent widths and colors and indicates an age of $\sim$ 500 Myr for these objects.

The Fritze v. Alvensleben & Kurth (1997) models can reproduce the H_β and H_γ equivalent widths without the need to invoke a non–standard IMF, although the agreement with H_δ is less good. The best agreement is with the 350 Myr model but the model continuum is redder than the observed spectra, even before correction for internal reddening (Brodie et al. 1997).

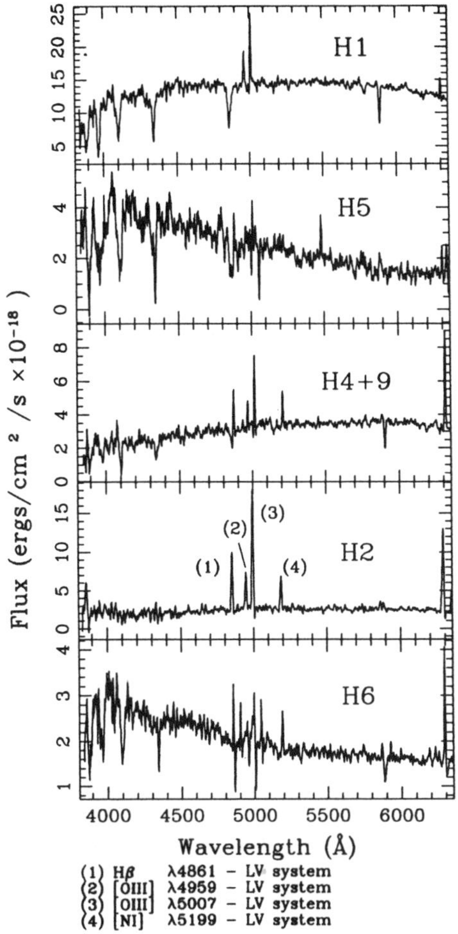

Figure 4. Fig. 4. – **Keck I spectra of NGC 1275 proto–globular cluster candidates.** The data have been smoothed with a box size of seven pixels and the resolution is 5.1 Å. The emission lines, which are residuals from imperfect subtraction of the background galaxy/gas spectrum, are marked to indicate the source of the line. LV: Residuals from the low–velocity system. HV: Residuals from the high–velocity infalling system.

The sense of the discrepancy is that the model predicts too much red light, consistent with the suggestion of a flatter cluster IMF.

Another problem with the assumption of a standard IMF for these objects is the fact that, based on their luminosities, the masses of these bright clusters are deduced to be $\sim 10^8 M_\odot$. Such high mass clusters would be difficult to form, requiring surprisingly massive progenitor gas clouds, and would be very unlikely to loose sufficient mass during their evolution to bring them into the mass range of normal old globular clusters. On the other hand, a pure A–star population, for example, would have a mass

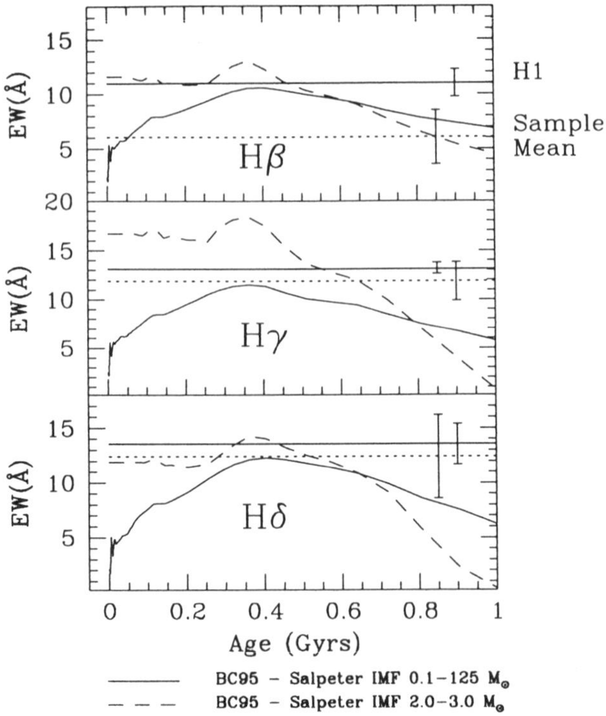

Figure 5. **Fig. 5.** – The time evolution of Balmer line equivalent widths for populations with different IMFs, produced by the Bruzual and Charlot models circulated in 1995. Also shown are the Balmer line equivalent widths measured for the brightest cluster, H1, (solid horizontal line) and the mean equivalent widths for the other clusters in the sample (dotted horizontal line). The value for Hβ is depressed, particularly in the lower signal-to-noise spectra, by under-subtracted Hβ emission in the galaxy background.

of $\sim 10^6 M_\odot$ at these luminosities, and a flatter than normal IMF would produce a mass somewhere in between.

Other key properties of the proto-globular cluster candidates are their spatial distribution and their velocity dispersion. These objects are extremely centrally concentrated. The entire sample (some 60 objects) is within 8 kpc of the nucleus and the brightest clusters are all within 2 kpc. We find a low velocity dispersion, $\sim$ 200 km/s, for our sample of 5 of the brighter candidates. This is lower than estimates of the stellar central velocity dispersion, $\sim$ 300 km/s (Lazareff et al. 1989).

3.1. CONSTRAINTS ON CLUSTER ORIGINS

The spectroscopic information allows us to set some interesting constraints on the origin of these objects. We can clearly rule out formation in a continuous cooling flow. The spatial scale of the clusters is very much less than the cooling flow radius (by a factor of 60 for our spectroscopic sample and a factor of 20 for the entire candidate sample) and their age is very much greater than the cooling time ($\sim$ 10 Myr). Moreover, a star formation rate in excess of 400 $M_\odot$/yr is deduced from the cooling flow (Allen & Fabian 1997). This implies a steep IMF because the far UV observations (Smith et al. 1992) set severe limits on the presence of high mass stars in this galaxy. We note that, if anything, the cluster's IMF may be biased *against* low mass stars.

It is equally clear that these clusters did not form in widespread shocks from merging galaxies. They are centrally concentrated and, if they formed far from the center and later fell in, a high rather than a low velocity dispersion would be expected.

It appears, then, that the clusters formed in a discrete event some 500 Myr ago. This may have been induced by a merger which provided the fuel for a short–lived gas inflow episode. It is not, however, clear to what extent processes such as these may have affected the global characteristics of globular cluster systems. The resultant clusters are not distributed like old globular clusters in central cD galaxies which are significantly more diffuse than the galaxy light. If they do indeed have an IMF which is biased against low mass stars, they may fade very rapidly. A pure A–star population would fade away in only $\sim 10^9$ yr.

4. Summary

- The observational evidence fails to match merger model expectations for globular cluster formation in several respects. A multi–phase collapse scenario may be more consistent with the observed properties of extragalactic globular cluster systems.
- The vast majority of globular clusters in NGC 1399 probably formed by the same processes and at the same epoch as those responsible for forming the Milky Way globular cluster system. Only a small subset (<5%) need have formed more recently (e.g. in a merger).
- The newly–defined V–I versus [Fe/H] relation suggests that metallicity differences between galaxies' globular cluster systems are much less significant than previously thought and that there are no very super–solar globular clusters.
- The proto–globular cluster candidates in NGC 1275 probably formed in a merger–induced gas infall episode some 500 Myr ago. However, their properties are such that they may not represent formation processes that had any significant effect on the global properties of globular cluster systems.

Acknowledgments

It is a pleasure to thank Duncan Forbes, Carl Grillmair, John Huchra, Markus Kissler-Patig, Drew Phillips and Linda Schroder who collaborated on the various parts of this work. This research was funded by HST grants GO.05990.01-94A and GO.05920.01-94A and by faculty research funds from the University of California, Santa Cruz.

REFERENCES

Ashman, K.M., & Zepf, S.E. 1992, ApJ, 384, 50
Allen, S.W., & Fabian, A.C. 1997, MNRAS, 286, 583
Brodie, J.P., & Huchra, J.P. 1990, ApJ, 362, 503
Brodie, J.P., & Huchra, J.P. 1991, ApJ, 379, 157
Brodie, J.P., Schroder, L.L., Huchra, J.P., Phillips, A.C., Kissler-Patig, M., & Forbes, D.A. 1997, AJ submitted
Bruzual, G., & Charlot, S. 1997, in preparation
Burstein, D., Faber, S.M., Gaskell, C.M., Krumm, N. 1984, ApJ, 287, 586
Fabian, A.C., Nulsen, P.E.J., & Canizares, C.R. 1984, Nature, 310, 733
Forbes, D., Brodie, J.P., & Grillmair, C. 1997, AJ, 113, 1652
Forbes, D.A., Grillmair, C.J., Williger, G.M., Elson, R.A.W., & Brodie, J.P. 1997, MNRAS, in press
Fritze-v. Alvensleben, U. & Burkert, A. 1995, A&A, 300, 58
Fritze-v. Alvensleben, U. & Kurth, O. 1997, in preparation
Harris, W.E. 1996, AJ, 112, 1487
Holtzman, J.A., et al. 1992, AJ, 103, 691
Kissler-Patig, M., Kohle, S., Hilker, M., Richtler, T., Infante, L., Quintana, H. 1997, A&A, 319, 470
Kissler-Patig, M. 1997, PhD thesis, Sternwarte, Bonn
Kissler-Patig, M., Kohle, S., Richtler, T., et al. 1997, A&A, 319, 470
Kissler-Patig, M., Brodie, J.P., Schroder, L.L., Forbes, D.A., Grillmair, C.G., & Huchra, J.P. 1998, AJ in press, January issue
Lazareff, B., Castets, A., Kim, D.W., & Jura, M. 1989, ApJ, 336, L13
Madore, B.F., Freedman, W.F. Silberman, N.A. et al. 1997, Nature, in press
Nørgaard-Nielsön, H.U., Goudfrooij, P., Jorgensen, H.E., & Hansen, L. 1993, AA, 279, 61
Ostrov, P., Geisler, D., Forte, J.C. 1993, AJ, 105, 1762
Rich, R.M., Sosin, G., Djorgovski, S.G., et al. 1997, ApJ, 484, L25
Richer, H., Crabtree, D., Fabian, A.., & Lin, D. 1993, AJ, 105, 877
Smith, E.P. et al. 1992, ApJ, 395, L49
Toomre A., Toomre J., 1972, ApJ 178, 623
Worthey, G., Faber, S.M., & Gonzales, J.J. 1992, ApJ 398, 69
Worthey, G. 1994, ApJS, 95, 107

THE EVOLUTION OF OLD STELLAR POPULATIONS IN OUR GALAXY

STEVEN R. MAJEWSKI
Department of Astronomy, University of Virginia
Charlottesville, VA 22903-0818, USA

1. Introduction

I would like to focus on one aspect regarding the evolution of Galactic stellar populations that is particularly relevant to discussions at this symposium: Where were the sites of early star formation in the Galaxy? The large scatter in abundance ratios for metal poor stars suggests multiple early settings of star formation in the Milky Way. In this and other ways, interpretation of detailed stellar chemical abundance analyses are converging with those of spatial-kinematical analyses of field stars, star clusters and satellite galaxies.

The latter now point to so-called "dual halo" models (Hartwick 1987, Zinn 1993, Majewski 1993, Norris 1994, Carney et al. 1996), which, in the simplest view, may be no more than a combination of the Eggen et al. (1962) collapse model leading to the formation of a flattened old population (most likely the Intermediate Population II – "IPII" – thick disk), with the Searle & Zinn (1978) scenario as the genesis of an "accreted" halo collected from disrupted satellites (see Sandage 1990). However, it would be imprudent not to state at the outset that the issue of the origin of the IPII remains unresolved, even at the most basic level of "top down" versus "bottom up" scenarios (cf. Majewski 1993).

Perhaps more progress is being made in understanding the origin of the more extended, outer halo. Evidence now suggests that this stellar population is dynamically unmixed – an important indication of continuing dynamical processes most likely related to accretion – and, while it dominates he most extreme spatial and kinematical distributions of the Milky Way, the outer halo does *not necessarily* dominate the most extreme [Fe/H] distribution in the Galaxy. Extremely metal poor stars are likely to make up both the accreted and flattened old populations of the Milky Way (Norris 1994). That the ages of some stars in the flattened halo are at

K. Nomoto and J.W. Truran (eds.), Cosmic Chemical Evolution, 185–193.

least as old as some in the outer halo, and perhaps even older (see below), suggests two major, parallel venues of star formation in the early Milky Way. A picture emerges whereby stars were forming in a major structure of substantially flattened, dissipated gas – leading to formation of the IPII and other flattened populations – nearly coincidentally with the onset of star formation in chemodynamically very different sites that gave birth to stars now constituting the outer halo. With an origin (possibly exclusively) through accretion, the outer halo represents stars and clusters likely to have formed in multiple, disconnected star formation sites (dwarf galaxies?) resulting in great chemical diversity. Depending on the precise nature of Galactic accretion processes, the ultimate number of original sites may still have been somewhat limited in the outer halo, perhaps in the form of several previously larger, LMC-like satellite galaxies.

2. Flattened, Old Stellar Populations

The need for a flattened, rotating, metal weak halo population, in addition to a more spherically shaped metal weak halo, is exhibited by RR Lyrae stars (Harwtick 1987), other horizontal branch (HB) stars (Kinman et al. 1994), dwarf stars (Sommer-Larsen & Zhen 1990, Ryan & Norris 1991, Allen et al. 1991, Majewski 1992, 1993, Carney et al. 1996, Wilhelm et al. 1997) and globular clusters (Zinn 1993). The fractional contributions of each of these subpopulations to the total "halo" remains to be clarified. The best that might be said is that in the local "halo" population, the densities of the two subpopulations are probably of the same order of magnitude (Sommer-Larsen & Zhen 1990, Kinman et al 1994). Understanding selection biases is crucial to this enterprise, and most critical is the distinction of "flattened halo" components from the IPII and other disk populations which overlap spatially and chemodynamically. In my opinion, such distinctions are obviated by our present inability to even identify distinguishing characteristics; so any separation represents an unnecessary obfuscation for the time being. With little weight of evidence suggesting otherwise, it is much simpler to consider all of the flattened, metal weak populations (e.g., the "metal weak thick disk", the "lower halo", the "flattened spheroid") as the same, and here I refer to this all-encompassing population as the IPII. Note that this simplification is not related to (the also unresolved) arguments over the separability of the IPII/thick and thin disks of the Milky Way – an issue with great bearing on the IPII origin (Gilmore et al. 1989, Majewski 1993).

What we *can* say with confidence is that the extreme age of stars and clusters in the flattened, metal-poor components are *old* – near the limits of Galactic stellar ages (Carney et al. 1990, Latham et al. 1992, Marquez & Schuster 1994, Gilmore et al. 1995, Agostinho et al. 1996, Schuster et al. 1996), and comparable to, if not older than, the outer, spherical halo subpopulation. This antiquity is supported by the few well-studied disk globular clusters (Fullton 1996), and of course by the "old halo" (à la Zinn 1993) globulars, but even the age distribution of *open clusters*, which canonically are a population of the *thin disk*, now extends past the ages of the youngest halo globular clusters (Phelps 1997; Majewski et al. 1996c, Figure 4). Relatively old ages ($>$ 12 Gyr) are also found for disk stars (Edvardsson et al. 1993, Nordstrom et al. 1997), and from white dwarf cooling theory (Isern et al. 1996, Oswalt et al. 1996). That some halo globular clusters (e.g., Rup 106, Pal 12, Ter 7, N6366, Arp 2) were forming gigayears after star formation in the disk commenced is strong evidence supporting late contributions to the outer halo after any global Galactic collapse (Searle & Zinn 1978). The combined distribution of open and globular cluster ages demonstrates the additional characteristic of a more or less continuous star formation rate in the integrated disk components, a fact also reflected in detailed chemodynamical analyses of starcount data (Haywood et al. 1997).

We must conclude that a primary site for early star formation was within the rather deep potential well of a fairly dissipated early Milky Way. Whatever the future direction of dynamical evolution of this flattened population (collapse with or without spin-up, heating by secular or stochastic processes), a signature of the *earliest* star formation here may be well-defined elemental abundance patterns specific to first nucleosynthetic sites of this particular environment (Brown et al. 1991, McWilliam et al. 1995) since there would be a lack of pre-enrichment by star formation sites now corresponding to the outer halo. Contributions by a possible pre-Galactic Population III (Ostriker, this symposium) would likely be rather homogeneous on relevant length scales. Observed disk age-[Fe/H] relations (Marsakov et al. 1990, Friel & Janes 1991, Edvardsson et al. 1993), while showing significant scatter at intermediate ages, do have more defined metal-poor tails at the greatest ages; however the latter must be regarded with great caution until survey selection biases are assessed. On the other hand, the remarkably small scatter in [α/Fe] in the low metallicity tail of disk stars (Edvardsson et al. 1993) attests to efficent large-scale mixing expected in the more global setting of this flattened halo, while the absolute value of the [O/Fe] ratio for these stars reflects enrichment dominated by Type II supernovae (Matteuci & Tornambé 1985, Truran & Thielemann 1987, Wheeler et al. 1989, Wyse & Gilmore 1988), consistent with expectations for a first population of Galactic stars. A practical complication to verification of a scenario as described is later contribution (via satellite mergers – a viable origin model for the "thick disk") of old stars formed in external sites to the flattened population of old stars formed *in situ*.

While the true identity of the damped Lyman α systems seen in high redshift QSOs is controversial, competing interpretations of these systems are particularly relevant to the dual halo picture of galaxy formation outlined here. Based on line profiles, Wolfe & Prochanska (1997) suggest that damped Ly α systems represent formation of galactic thick disks as described above. On the other hand, abundance patterns suggest a closer association of damped Ly α systems to "halo clouds", reminiscent of dwarfish galaxies (Lu, this symposium), perhaps like the Small Magellanic Cloud (York, this symposium). These "halo clouds" in high redshift systems may be related to accreted systems now evident in our own Galaxy.

3. Accretion in the Halo

Cold dark matter cosmological models predict large amounts of accretion by galaxies since $z \approx 3$, usually by the accumulation of discrete, "sub-galactic units". In "standard" CDM, the majority of dark halos have endured $\approx 10\%$ mass mergers since $z = 0.4$ (Frenk et al. 1988, Kauffman & White 1993, Lacey & Cole 1993), and since then their embedded L^* galaxies sustained smaller, yet still significant merger rates (Carlberg 1990, Tóth & Ostriker 1992, Navarro et al. 1994). Assessments of the rate of decay of satellite orbits from dynamical friction in dark matter halos also point to the normalcy of such events in the lives of Milky Way-like galaxies, to the level of an LMC mass per Hubble time (Ostriker & Tremaine 1975, Hernquist 1991, Mihos & Bothun 1997). Thus, it would be peculiar *not* to identify a significant role for accretion in the evolution of our Galaxy.

First strong clues that the outer halo endured a protracted phase of accumulation and integration of initially chemodynamically independent systems derived from analysis of the globular cluster system (Searle & Zinn 1978). More recent analysis of the globular system incorporating kinematics, spatial distributions, [Fe/H] and HB morphology (Zinn 1993, van den Bergh 1993), nicely demonstrates the utility of a "dual halo" model (Zinn 1993, Da Costa & Armandroff 1995). If the second parameter of HB morphology is age, a slow formation – at least 3 Gyr – is suggested for the spherically distibuted, "young halo" in which reside the youngest (i.e., "second parameter") halo globular clusters. This spread of ages is robust to interpretations of the second parameter for at least some clusters dated by other means, such as main sequence turn offs (see summary in Richer et al. 1996). The "old halo" clusters are in a flattened, rotating distribution and, when considered in combination with the disk globular clusters, exhibit ELS "spin-up". The kinematics of the "young halo" system, the lack of a metallicity gradient therein, and the apparent relatively younger age for this population compared to the "old halo" + disk system all support the original Searle & Zinn scenario of accumulation of "fragments" in the outer halo even after the flattened, "old halo" system was well established. Evidence for association of specific halo clusters with Galactic satellite galaxies – Ter 7, Arp 2, Ter 8 and M54 with Sagittarius (Ibata et al. 1995) and Pal 12 and Rup 106 with the Magellanic Clouds (Lin & Richer 1992) strengthens the notion, anticipated by Rodgers & Paltoglou (1984), that a number of outer halo globulars may have been accreted from dwarf satellites of the Milky Way. The majority of these suspect clusters appear to belong to Zinn's "young halo" and, statistical correlations of extreme second parameter clusters with dwarf galaxies (Lynden-Bell 1982, Majewski 1994, Lynden-Bell & Lynden-Bell 1995, Fusi Pecci et al. 1995) provide a natural explanation for the ages (Majewski 1994), though exceptions (e.g., Ter 8) do muddy the picture somewhat (Da Costa & Armandroff 1995). Of course, there is still no concensus that age *is* the sole second parameter (Catelan & de Freitas Pacheco 1996, Buonanno et al. 1997, Ferraro et al. 1997) although this controversy in itself does not preclude accretion as an explanation for HB differences, whatever the direct physical cause.

The age scatter in halo field stars is at least as great as that in the young halo cluster system (Schuster et al. 1996), and perhaps greater – as shown by the apparently young, blue metal poor stars of Preston et al. (1994) and the carbon stars of Totten & Irwin (1996), both of which are attributed to satellite accretion. Evidence for significant phase space complexity, manifested as "moving groups", streams, or other kinematic substructure of halo stars (Carney et al. 1996, Majewski et al. 1996a, summary in Majewski et al. 1996b) is consistent with a dynamically unrelaxed, accreted outer halo.

Direct evidence that halo accretion may be a ubiquitous phenomenon and provide an important source for stars in the "outer halo" is of course provided by the Sgr dwarf galaxy (Ibata et al. 1995). Sgr has now been shown to extend to some 40° in length (Majewski et al. 1997a), but, according to models by Johnston (1997), it would take only 3 Gyr for this debris to circle all of the way around the Galaxy, and in 10 Gyr Sgr stellar debris may cover a substantial fraction of the sky (15% if multiple wrappings overlap, and more if not). But this is factors smaller than the model results for sky coverage of debris from the break-up of the Magellanic Clouds after 10 Gyr. The latter prediction is in apparent contradiction with failed searches for excesses of main sequence stars (Mathewson et al. 1979, Recillas-Cruz 1982, Brück & Hawkins 1983, Guhathakurta & Lin 1997) near the H I Magellanic Stream.

However, we (Majewski et al. 1997b) have been conducting a survey for *giant stars* in a 45° swath cutting across the H I stream some 20° downststream from the MC center and have found some provocative results consistent with Johnston's models. Giant stars, while not as abundant as main sequence stars, are much easier to find since they may be identified photometrically (with intermediate band filters) in large area, shallow surveys. While a spectroscopically verified sample of giants to $V \approx 18$ indicates a more or less even sky density across the ten fields of ~ 4 deg^{-2}, the giants show some correlations between apparent distance and longitude. For example, a group of giants, found mainly in the $l = 300°$ to $315°$ fields, have a distance grouped at ≈ 25 kpc, while most of the giants between $l = 275°$ and $290°$ are grouped at a distance some 2.5 times greater. To some degree, these data match predictions of tidal stream locations in the Galactic halo derived from new semi-analytical models by K. Johnston of the tidal disruption of the LMC over the last 10 Gyr. In these models, the multiply wrapped Toomre bridge streamer concentrates stars to the range 20-40 kpc, while the Toomre tail stars tend to concentrate at larger distances, 50-80 kpc. Precessional shifts of inner and outer concentrations are roughly consistent with the apparent longitudinal concentrations found in our giant sample. This general correspondence of the model predictions with the early results of our survey suggests the possibility that we may be identifying *widely dispersed* tidal star streams of the Magellanic Clouds.

Disruption of the Magellanic Clouds was not accounted for in the analysis of the amount of accretion in the halo by Unavane et al. (1996), on the grounds that the mean Cloud metallicity today is much too high compared to the bulk of the halo field stars. They also point out the significantly higher mean abundance of the present retinue of Galactic dSph's – when luminosity weighted – over that of the halo field stars, and they conclude that (1) $< 3\%$ of the halo could have derived from Carina-like predecessors (a galaxy which contains significant numbers of intermediate aged stars; Smecker-Hane et al. 1994), and (2) the fraction of halo stars younger than the dominant halo population is $< 8\%$ for [Fe/H]< -1.5 – consistent with estimates from the blue metal-poor stars of Preston et al. (1994). However, substantial *early* shredding of dwarf galaxies which are no longer, or only marginally, represented in a luminosity-weighted distribution (e.g., Ursa Minor, Draco?) of the present Galactic satellite family, is not ruled out (as they point out). Moreover, if the Magellanic Clouds have been disintegrating for nearly a Hubble time, the most widely dispersed stars would presumably be the oldest and most metal-poor; indeed, the oldest, most metal-poor stars in the LMC are found to be as old as those in the Milky Way (Barbuy, this symposium).

The implications of continuous shredding (from early times) of substantially larger satellites, like the Magellanic Clouds, Sgr and Fornax have yet to be fully worked out. But the correlation of the present dSph's and young halo clusters in the sky suggest several possible accretionary paths. Dwarf spheroidals *themselves* may have derived from larger dIrr galaxies, possibly as a first phase of break up of the latter (Lynden-Bell 1982, 1994, Kroupa

1997), followed by a second phase whereby the daughter dSph break up to contribute further to the halo (Kuhn et al. 1997). There is utility to this evolutionary path for explaining multiple generations of stars in the dSph (Smecker-Hane et al. 1994), which presents a somewhat problematical dilemma if these dwarf systems are not sufficiently massive to retain gas after starbursts; initial populations in these systems may have formed before breakoff from a larger system. Galaxy interactions involving larger satellites are seen to *initiate* formation of both globular clusters (Ashman & Zepf 1992, Schweizer et al. 1996, Brodie, this symposium) *and* dwarf galaxies (Mirabel et al. 1992), showing that these objects might constitute from gas (and stars) derived from larger dIrr galaxies, but external to them. The young, blue star associations in the intercloud region of the Magellanic system (Kunkel 1980, Irwin et al 1990) may present us with a more local paradigm for such a phenomenon. The lumpy, and widespread distribution of high velocity H I (Murphy et al. 1995) may be a signature of gas stripping from large satellites, as the H I Magellanic Stream almost certainly is. The high velocity clouds have low metallicity, but this is highly variable from complex to complex. The general conclusion is that this gas cannot be *entirely* primordial (Schwarz et al. 1995, Wakker 1991)

If accretion of chemically individuated stellar sytems plays a key role in the formation of the outer halo, one might expect rather chaotic chemical abundance patterns for stars found there. Depending on star formation histories and dynamical disruption timescales for individual progenitors, abundance patterns from accreted stars will reflect differing ratios of Type II or Type Ia supernovae products. However, if the filling factor of stellar streams is as low as suggested by Majewski et al. (1996), then local volumes of the accreted halo may be dominated by stars from a small number of progenitor star formation sites. Therefore, study of abundance patterns in outer halo stars will provide important complementary fossil data to halo phase space distributions as a means to unravel the Galaxy's merging history. Nice demonstrations of the bright future in the industry of combining detailed chemical and kinematical abundances are (1) the finding (Brown et al. 1997) of low [O/Fe] ratios (compared to most halo stars), characteristic of enrichment by Type Ia supernovae in an environment with multiple star formation events and similar to those in the Magellanic Clouds (Gilmore & Wyse 1991), in the the rogue, young clusters Rup 106 and Pal 12, suspected from their kinematics of being from the Magellanic Clouds (Lin & Richer 1992); (2) King's (1997) finding of comparably low [α/Fe] in high proper motion stars with kinematics suggesting accretion; and (3) Nissen & Schuster's (1997) demonstration that the lowest [α/Fe] and [Na/Fe] are found in stars having kinematics yielding the largest R_{max} and Z_{max} distances.

References

Agostinho, R., Anton, S., Maia, D., Rose, J. & Stock, J. 1996, in *Formation of the Galactic Halo...Inside and Out*, eds. H. Morrison & A. Sarajedini, ASP Conf. Ser. 92, 65

Allen, C., Schuster, W. J. & Poveda, A. 1991, *A&A*, **244**, 280

Ashman, K. M. & Zepf, S. E. 1992, *ApJ*, **384**, 50

Brown, J. A., Wallerstein, G. & Zucker, D. 1997, *AJ*, **114**, 180

Brown, J. H., Burkert, A. & Truran, J. W. 1991, *ApJ*, **376**, 115

Buonnano, R., Corsi, C., Bellazini, M., Ferraro, F. & Fusi Pecci, F. 1997, *AJ*, **113**, 706

Brück, M. T. & Hawkins, M. R. S. 1983, *A&A*, **124**, 216

Carlberg, R. G. 1990, *ApJ*, **359**, L1

Carney, B. W., Latham, D. W. & Laird, J. B. 1990, *AJ*, **99**, 572

Carney, B.W., Laird, J. B., Latham, D. W. & Aguilar, L. A. 1996, *AJ*, **112**, 668

Catelan, M. & de Freitas Pacheco, J. A., 1995, *A&A*, **297**, 345

Da Costa, G. S. & Armandroff, T. E. 1995, *AJ*, **109**, 2533

Edvardsson, B., Andersen, J., Gustafsson, B., Lambert, D. L., Nissen, P. E. & Tomkin, J. 1993, *A&A*, **275**, 101

Eggen, O. J., Lynden-Bell, D. & Sandage, A. R. 1962, *ApJ*, **136**, 748

Ferraro, F. R., Paltrinieri, B., Fusi Pecci, F., Cacciari, C.., Dorman, B. & Rood, R. T. 1997, *ApJ*, **484L**, 145

Frenk, C., White, S. D. M., Davis, M. & Efstathiou, GH. 1988, *ApJ*, **327**, 507

Friel, E. D. & Janes, K. A. 1991, in *The Formation and Evolution of Star Clusters*, ed. K. Janes, ASP Conf. Ser. 13, 569

Fullton, L. K. 1996, *PASP*, **108**, 545

Fusi Pecci, F., Bellazzini, M., Cacciari, C. & Ferraro, F. R. 1995, *AJ*, **110**, 1664

Gilmore, G. & Wyse, R. F. G. 1991, *ApJ*, **367**, L55

Gilmore, G., Wyse, R.F. G. & Kuijken, K. 1989. *ARAA*, **27**, 555

Gilmore, G., Wyse, R. F. G. & Jones, B. J. 1995, *AJ*, **109**, 1095

Guhathakurta, P. & Lin, D. N. C. 1997, *in preparation.*

Hartwick, F. 1987, in *The Galaxy*, eds. G. Gilmore & B. Carswell, (Reidel: Dordrecht), p.281

Haywood, M., Robin, A. C. & Creze, M. 1997, *A&A*, **320**, 440

Hernquist, L. 1991, in *Warped Disks and Inclined Rings Around Galaxies*, eds. S. Casertano, P. Sackett & F. Briggs, (Cambridge: Cambridge Univ. Press, p. 96

Ibata, R. A., Gilmore, G. & Irwin, M. J. 1995, *MNRAS*, **277**, 781

Irwin, M. J., Demers, S. & Kunkel, W. E. 1990, *AJ*, **99**, 191

Isern, J., Hernanz, M., Garciá-Berro, Mochkovitch, R. & Burkert, A. 1996, *The History of the Milky Way and Its Satellite System*, ASP Conf. Ser. 112, eds. A. Burkert, D. H. Hartmann & S. R. Majewski, (ASP: San Francisco), p. 181

Johnston, K. V. 1997, *ApJ*, *in press.*

Kauffman, G. & White, S. D. M. 1993, *MNRAS*, **261**, 921

King, J. 1997, *AJ*, **113**, 2302

Kinman, T. D., Suntzeff, N. B. & Kraft, R. P. 1994, *AJ*, **108**, 1722

Kroupa, P. 1997, *NA*, **2**, 139

Kuhn, J. R., Smith, H. A. & Hawley, S. L. 1996, *ApJLett*, 469, L93

Kunkel, W. E. 1979, *ApJ*, **228**, 718

Kunkel, W. E. 1980, in *Star Clusters*, ed. J. Hesser, (Dordrecht: Reidel), p.353

Lacey, C. & Cole, S. 1993, *MNRAS*, **262**, 627

Latham, D. W., et al. 1992, *AJ*, **104**, 774

Lin, D. N. & Richer, H. B. 1992, *ApJ*, **388**, L57

Lynden-Bell, D. 1982, *Observatory*, **102**, 202

Lynden-Bell, D. 1994, in *Dwarf Galaxies*, eds. G. Meylan & P. Prugniel, (ESO), p.589

Lynden-Bell, D. & Lynden-Bell, R. M. 1995, *MNRAS*, **275**, 429

Majewski, S. R. 1992, *ApJSupp*, **78**, 87

Majewski, S. R. 1993, *ARAA*, 31, 575

Majewski, S. R. 1994, *ApJLett*, **431**, L17
Majewski, S. R., Hawley, S. L. & Munn, J. A. 1996b, in *Formation of the Galactic Halo...Inside and Out*, eds. H. Morrison & A. Sarajedini, ASP Conf. Ser. 92, 119
Majewski, S. R., Munn, J. A. & Hawley, S. L. 1996a, *ApJLett*, **459**, L73
Majewski, S. R., Phelps, R. & Rich, R. M. 1996c, in *The History of the Milky Way and Its Satellite System*, eds. A. Burkert, et al., ASP Conf. Ser. 112, 1.
Majewski, S. R., Ostheimer, J. C., Kunkel, W. E. & Johnston, K. V. 1997b, *in preparation*
Majewski, S. R., Siegel, M. H., Reid, I. N., Thompson, I. B., Landolt, A. & Kunkel, W. E. 1997a, *ApJLett, submitted.*
Marquez, A. & Schuster, W. J. 1994, *A&AS*, **108**, 341
Marsakov, V. A., Shevelev, Y. G. & Suchkov, A. A. 1990, *Astrophys. Space Sci.*, **172**, 51
Mathewson, D. S., Ford, V. L., Schwarz, M. P. & Murray, J. D. 1979, in *The Large-Scale Characteristics of the Galaxy*, ed. W. B. Burton, (Dordrecht: Reidel), p. 547
Matteuci, F. & Tornambé, A. 1985, *A&A*, **142**, 13
McWilliam, A., Preston, G. W., Sneden, C. & Searle, L. 1995, *AJ*, **109**, 2757
Mihos, J. C. & Bothun, G. D. 1997, *ApJ*, **481**, 741.
Mirabel, I. F., Dottori, H. & Lutz, D. 1992, *A&A*, **256**, L19
Murphy, E. M., Lockman, F. J. & Savage, B. D. 1995, *ApJ*, **447**, 642
Navarro, J. F., Frenk, C. S. & White, S. D. M. 1994, *MNRAS*, **267L**, 1
Nissen, P. E. & Schuster, W. J. 1997, *A&A*, **326**, 751
Nordström, B., Olsen, E. H., Andersen, J., Mayor, M. & Pont, F. 1997, in *The History of the Milky Way and Its Satellite System*, ASP Conf. Ser. 112, eds. A. Burkert, D. H. Hartmann & S. R. Majewski, (ASP: San Francisco), p. 145
Norris, J. 1994, *ApJ*, **431**, 645
Ostriker, J. P. & Tremaine, S. 1975, *ApJ*, **202**, L113
Oswalt, T. D., Smith, J. A., Wood, M. A. & Hintzen, P. 1996, *Nature*, **382**, 6920
Phelps, R. L. 1997, *ApJ*, **483**, 826
Preston, G. W., Beers, T. C. & Shectman, S. A. 1994, *AJ*, **108**, 538
Recillas-Cruz, E. 1982, *A&A*, **124**, 216
Richer, H. B., et al. 1996, *ApJ*, **463**, 602
Rodgers, A. W. & Paltoglou, G. 1984, *ApJ*, **283**, L5
Ryan, S. G. & Norris, J. E. 1991, *AJ*, **101**, 1835
Sandage, A. 1990, *JRASC*, **84**, 70
Schuster, W., Nissen, P., Parrao, L., Beers, T. & Overgaard, L. 1996, *A&AS*, **117**, 317
Schwarz, U. J., Wakker, B. P. & van Woerden, H. 1995, *A&A*, **302**, 364
Schweizer, F., Miller, B. W., Whitmore, B. C. & Fall, S.M. 1996, *AJ*, **112**, 1839
Searle, L. & Zinn, R. 1978, *ApJ*, **225**, 357
Smecker-Hane, T. A., Stetson, P. B., Hesser, J. E. & Lehnert, M. D. 1994, *AJ*, **108**, 507
Sommer-Larsen, J. & Zhen, C. 1990, *MNRAS*, **242**, 10
Tóth, G. & Ostriker, J. P. 1992, *ApJ*, **389**, 5
Totten, E. J. & Irwin, M. J. 1996, in *Formation of the Galactic Halo...Inside and Out*, eds. H. Morrison & A. Sarajedini, ASP Conf. Ser. 92, 200
Truran, J. W. & Thielemann, F. K. 1987, in *Stellar Populations*, ed. C. A. Norman, A. Renzini & M. Tosi, (Cambridge: Cambridge Univ. Press), p. 149
Unavane, M., Wyse, R. F. G. & Gilmore, G. 1996, *MNRAS*, **278**, 727
van den Bergh, S. 1993, *AJ*, **105**, 971
Wakker, B. P. 1991, in *IAU Symposium 144: The Interstellar Disk-Halo Connection in Galaxies*, ed. H. Bloemen, (Kluwer: Dordrecht), p. 27
Wheeler, J. C., Sneden, C. & Truran, J. W. 1989, *ARAA*, **27**, 279
Wilhelm, R., Beers, T. C., Kriessler, J. R., Pier, J. R., Sommer-Larsen, J. & Layden, A. C. 1997, *preprint*
Wolfe, A. & Prochanska, J. X. 1997, *preprint*
Wyse, R. F. G. & Gilmore, G. 1988, *AJ*, **95**, 1404
Zinn, R. 1993, in *The Globular Cluster – Galaxy Connection*, ASP Conf. Ser. 48, eds. Smith, G. & Brodie, J., (ASP: San Francisco), p. 38

CHEMICAL EVOLUTION OF THE MAGELLANIC CLOUDS

B. BARBUY
Universidade de São Paulo
C.P. 3386, São Paulo 01060-970, Brazil

J. A. DE FREITAS PACHECO
Observatoire de la Côte d'Azur
BP 4229, 06304 Nice Cedex 4, France

AND

T. IDIART
Observatoire de la Côte d'Azur
BP 4229, 06304 Nice Cedex 4, France

1. Introduction.

Chemical evolution studies of the Clouds based on detailed abundance analyses are essentially limited nowadays to young objects like HII regions (Dufour 1984; Garnett et al. 1995), planetary nebulae (e.g. de Freitas Pacheco et al. 1993), supernova remnants (Russell & Dopita 1990) and supergiant stars either of F-K type (Hill et al. 1995) or B ones (e.g. Rolleston et al. 1996 and references therein). One exception to this was presented by Dopita et al. (1997) where HST data permitted to date planetary nebulae nuclei and to present an age vs. metallicity model.

Otherwise, high-resolution abundance analyses of individual stars of old populations are awaiting for the availability of 8m telescopes in the southern hemisphere, in which case stars of V > 16 will be feasible.

The large number of star clusters in the Clouds make them interesting probes to check chemical evolution models of these nearby galaxies (Olszewski et al. 1991), since these clusters have ages between a few million years up to 14 Gyr. Many of these clusters have ages estimated from colour-magnitude diagrams (CMD), and metallicities derived from analyses of individual stars.

In the present work we apply SSP models that we have developed in the past few years to MC clusters (Borges et al. 1995). The accuracy of our age and

K. Nomoto and J.W. Truran (eds.), Cosmic Chemical Evolution, 195–199.

metallicity determinations based on integrated indices is checked against well-determined ages obtained with Colour-Magnitude Diagrams (CMDs) and metallicities derived from spectroscopy of individual stars. This is an ongoing project, where for the moment we have observed an initial sample of 14 objects amongst those most well studied clusters in the Clouds. These well-known clusters are used to check our procedure, which will be applied in the future to unstudied clusters.
Besides, a chemical evolution model is also proposed, where a star formation rate based on information from the field population is adopted. The model is compared to our star clusters data.

2. Studies of field populations

The old populations can be studied either through the field or the star cluster populations.
Recently, great attention is given to the field populations. Through imaging and extraction of colour-magnitude diagrams of field populations in the LMC mainly with HST, but also with NTT, a number of authors (Bertelli et al. 1992; Ardeberg et al. 1997; Westerlund et al. 1995; Gallagher et al. 1996; Vallenari et al. 1996; Elson et al. 1997; Stappers et al. 1997) have found essentially 3 stellar populations: an old population covering the age range $5 < \tau < 14$ Gyrs, an intermediate age of $2 < \tau < 4$ Gyrs, and a young population ($\tau < 1$ Gyr) near the LMC bar.

These data can be explained by an initial steady star formation rather than a star formation history dominated by intense bursts, as previously assumed in the literature. One burst would have occurred at 2-4 Gyrs, when about 30% of the total number of stars were formed.

3. Star clusters

The star clusters are bright suitable tracers, and they are numerous in the Clouds. However, for the LMC there are no clusters in the metallicity range -1.8 < [Fe/H] < -0.7, which is equivalent to a gap in the age interval $4 < \tau < 12$ Gyr. In the SMC there are clusters at all metallicities and ages. The age and metallicity distribution in the LMC has misled authors, until recently, as concerns its star formation history: a history of bursts had been favoured. In fact, due to the metallicity (age) gap, clusters can be used as tracers of chemical evolution but not of star formation rate. For the star formation rate, the data on field stars, now available, are appropriate.
Most studies of clusters were based on integrated photometry (Searle et al. 1980; Frenk & Fall 1982; Elson & Fall 1988; Girardi & Bica 1993; Girardi et al. 1995). These authors have shown that the age gap is difficult to detect in colours since, for example Δ(U-B) = 0.03 for ages between 12 and 4 Gyr.

It is interesting to note that essentially no studies were carried out based on integrated spectroscopy, with the exception of that by Bica et al. (1990) for young clusters ($\tau < 0.5$ Gyr). The present work consists of observations and interpretation of integrated spectra of star clusters in both Clouds.

4. Present method

An initial sample of 6 clusters in the SMC and 8 ones in the LMC were respectively observed at the *Laboratório Nacional de Astrofísica* (LNA), Brazil and at the *European Southern Observatory* (ESO), Chile. This is an ongoing project, and observations of a much larger sample are planned.
The indices Hβ, Mg_2, Fe5270, Fe5335 and NaD were measured using the continuum and central bandpasses defined by Faber et al. (1985).
Using relations between <Fe>, Hβ vs. stellar parameters (T_{eff}, log g, [Fe/H], [α/Fe]) integrated indices for single stellar populations (SSPs) were build up, using Padova evolutionary tracks and a Salpeter's law for the IMF.

For <Fe> and Hβ indices, and (B-V) colours, the relations for SSPs (where ages are given in Gyrs) are from Borges et al. (1995):

$$\langle Fe \rangle = 2.319 + 0.283 \ln \tau + 1.968[Fe/H] + 0.393[Fe/H]^2$$

$$H\beta = \exp(1.219 - 0.343 \ln \tau - 0.344[Fe/H] + 0.015[Fe/H]*\ln \tau$$

$$(B\text{-}V)_o = 0.726 + 0.091 \ln \tau + 0.269[Fe/H] - 0.032[Fe/H]*\ln \tau$$

In the present work the colour $(V - K_{CIT})$ was synthesized using the same procedure, adopting the temperature scale, $\log g$ and $(V - K_{CIT})_0$ from the work by Frogel et al. (1983), since the $(V - K_{CIT})$ data for the MC clusters by Persson et al. (1983) were obtained in the same photometric system. The following relation is obtained:

$$(V\text{-}K_{CIT}) = 2.676 + 0.180 \ln \tau + 0.547[Fe/H]$$

5. Chemical evolution models

The models consider yields by Nomoto et al. (1984) for C-deflagration models of SNIa (10-60$M_\odot$), and from Thielemann et al. (1996) for SNII.
The adopted constraints for the LMC are: a present fraction of gas $f_{gas}^{present}$ = 0.11, a present star formation rate $SFR^{present}$ = 0.12 $M_\odot$/yr and the present oxygen-to-iron ratio of $[O/Fe]^{present}$ = -0.2 (Hill et al. 1997). For the SMC the constraints are: $f_{gas}^{present}$ = 0.33 (Lequeux 1993), SFR $\propto M_{gas}$, $SFR^{present}$ = 0.052 $M_\odot$/yr.

Based on a star formation history as indicated by the field population (Sect. 2), the age-metallicity relation for the LMC is shown in Fig. 1, where our sample clusters are also plotted.

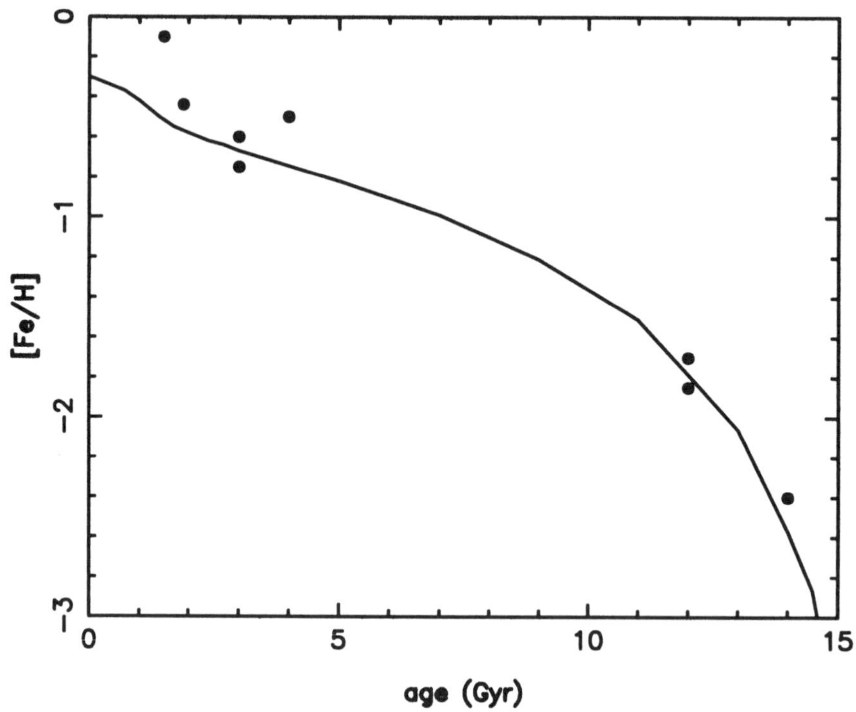

Figure 1. **LMC: Chemical evolution model (solid line) compared to age and metallicity data for the present sample clusters** (NGC 1466, NGC 1783, NGC 1795, NGC 1978, NGC 2121, NGC 2173, NGC 2210, Hodge 11)

6. Conclusions

We have obtained integrated spectra for 14 clusters in the Magellanic Clouds, on which the spectral indices Hβ, Mg_2, Fe5270, Fe5335 were measured.

Selecting indices whose behaviour depends essentially on age and metallicity (Hβ and $< Fe >$), together with (B-V) and (V-K) colours, we were able to determine age and metallicities for these clusters, using calibrations based on single stellar population models (Borges et al. 1995).
A chemical evolution model which follows a star formation history as indicated by the field population is checked with the age and metallicity data for our sample star clusters.

References

Ardeberg, A., Gustafsson, B., Nissen, P.-E. 1997 A&A, 322, L13

Bertelli, G., Mateo, M., Chiosi, C., Bressan, A. 1992, ApJ, 388, 400
Bica, E., Alloin, D., Santos Jr., J.F.C. 1990: A&A, 235, 103
Borges, A.C., Idiart, T.P., de Freitas Pacheco, J. A., & Thevenin, F. 1995, AJ, 110, 2408
Dopita et al. 1997, ApJS, 108, 515
Dufour, R.J. 1984, in *Structure and evolution of the Magellanic Clouds*, IAU Symp. 108, eds. S. van den Bergh, K. de Boers, Reidel (Dordrecht), p. 353
Elson, R., Fall, M. 1985, ApJ, 299, 211
Elson, R., Gilmore, G., Santiago, B. 1997, MNRAS, in press
Faber, S.M., Friel, E.D., Burstein, D., & Gaskell, C.M. 1985, ApJS 57, 711
Frenk, C., Fall, M. 1982, MNRAS, 199, 565
de Freitas Pacheco, J.A., Barbuy, B., Costa, R.D.D., & Idiart, T. 1993, A&A, 271, 429
Frogel, J., Persson, S.E., Cohen, J.G. 1983, ApJS, 53, 71
de Freitas Pacheco, J.A. 1997, A&A, 319, 394
Gallagher et al. 1996, ApJ, 466, 732
Garnett, D.R., Skillman, E.D., Dufour, R.J., Peimbert, M., Torres-Peimbert, S., Terlevich, R., Terlevich, E., & Shields, G.A. 1995, ApJ, 443, 64
Girardi, L., Bica, E. 1993, A&A 274, 279
Girardi, L., Chiosi, C., Bertelli, G., Bressan, A. 1995, A&A 298, 87
Hill, V., Andrievsky, S., & Spite, M. 1995, A&A, 293, 347
Hill, V., Barbuy, B., Spite, M. 1997, A&A, 323, 461
Lequeux, J. 1993, ESO/OHP workshop on Dwarf Galaxies, eds. G. Meylan & P. Prugniel, ESO/CNRS, p. 179
Nomoto, K., Thielemann, F.-K., Yokoi, K. 1984, ApJ, 286, 644
Olszewski, E.W., Schommer, R.A., Suntzeff, N., & Harris, H.C. 1991, AJ, 101, 515
Persson, S.E., Aaronson, M., Cohen, J.G., Frogel, J.A., & Matthews, K. 1983, ApJ, 266, 105
Rolleston, W., Brown, P., Dufton, P., Howarth, I. 1996, A&A, 315, 95
Russell, S.C., Dopita, M.A. 1990, ApJ, 384, 508
Stappers et al. 1997, PASP, 109, 292
Thielemann, F.-K., Nomoto, K., Hashimoto, M. 1996, ApJ, 460, 408
Vallenari, A., Chiosi, C., Bertelli, G., Ortolani, S. 1996, A&A, 309, 358
Westerlund, B., Linde, P., Lynga, G. 1995, A&A, 298, 39

GLOBAL EVOLUTION OF THE STARS, GAS, METALS, AND DUST IN GALAXIES

S. MICHAEL FALL
Space Telescope Science Institute
3700 San Martin Drive, Baltimore, MD 21218, USA

Abstract. We outline a method to infer the global history of star formation in galaxies from absorption-line observations of quasars. The application to existing data leads to the conclusion that most stars formed at relatively low redshifts ($z \lesssim 2$). The results obtained by this absorption-based method are consistent with those obtained subsequently by emission-based methods.

1. Introduction

In this article, we consider the evolution of, and relations between, various large-scale average properties of the population of galaxies as a whole. It is often convenient to express these "global" properties as mean comoving densities and to normalize them to the present closure density. We are particularly interested in the comoving densities of stars, gas, metals, and dust within galaxies, which we denote respectively by Ω_s, Ω_g, Ω_m, and Ω_d. The last three of these are meant to refer to the interstellar media (ISM) of galaxies, exclusive of the intergalactic medium (IGM), although in practice such a distinction may only be approximate. As defined here, Ω_m includes metals in both the gas and solid (i.e., dust) phases of the ISM. It is usually more informative to reexpress Ω_m and Ω_d in terms of the mean metallicity, $Z \equiv \Omega_m/\Omega_g$, and the mean dust-to-gas ratio, $D/G \equiv \Omega_d/\Omega_g$. It is clear that all of these properties are related in the sense that, as new stars form, Ω_s will increase, while in most cases, Ω_g will decrease and Z and D/G will increase. The goal here is to quantify such notions through the equations of "cosmic chemical evolution."

Until recently, there were no emission-based estimates of the global rate of star formation $\dot{\Omega}_s$ at $z \gtrsim 0.3$. The reason for this is that samples of galaxies selected by emission become progressively incomplete and include

K. Nomoto and J.W. Truran (eds.), Cosmic Chemical Evolution, 201–207.

only brighter objects at higher redshifts. In contrast, samples of galaxies selected by absorption against background quasars do not suffer from this bias. Such observations are exquisitely sensitive to small column densities of absorbing or scattering particles. In principle at least, they enable us to estimate Ω_g, Ω_m, and even Ω_d as functions of redshift. From these and the equations of cosmic chemical evolution, we can then infer the global rate of star formation $\dot{\Omega}_s$. It is amusing to note that this idealistic program does not require the detection of a single stellar photon! We can also combine our estimates of $\dot{\Omega}_s$ with stellar population synthesis models to compute the mean comoving emissivity $\mathcal{E}_\nu$ and the mean intensity of background radiation J_ν. One might then claim to have predicted the "emission history" of the universe from its "absorption history." This article describes a first attempt by Yichuan Pei, Stéphane Charlot, and the author to carry out such a program; a complete account of our work is given elsewhere (Pei & Fall 1995; Fall et al. 1996).

2. Damped Lyman-Alpha Galaxies

Before proceeding, it is worth recalling some facts about the statistics of absorption-line systems. Let $f(N_x, z)$ be the column density distribution of particles of any type x that absorb or scatter light. These might, for example, be hydrogen atoms ($x =$ HI), metal ions ($x = m$), or dust grains ($x = d$). By definition, $H_0(1+z)^3|dt/dz|f(N_x, z)dN_x dz$ is the mean number of absorption-line systems with column densities of x between N_x and $N_x + dN_x$ and redshifts between z and $z+dz$ along the lines of sight to randomly selected background quasars. These lines of sight are very narrow (less than a light year across) and pierce the absorption-line systems at random angles and impact parameters. One can show that the mean comoving density of x is given by

$$\Omega_x(z) = \frac{8\pi G m_x}{3cH_0} \int_0^\infty dN_x f(N_x, z) N_x, \tag{1}$$

where m_x is the mass of a single particle (atom, ion, or grain). Equation (1) plays a central role in this subject. It enables us to estimate the mean comoving densities of many quantities of interest without knowing anything about the structure of the absorption-line systems. In particular, we do not need to know their sizes or shapes, whether they are smooth or clumpy, and so forth. A corollary of equation (1) is that the global metallicity, $Z \equiv \Omega_m/\Omega_g$, is given simply by an average over the metallicities of individual absorption-line systems weighted by their gas column densities.

The absorption-line systems of most interest in the present context are the damped Lyα (DLA) systems. It is widely believed that they trace the ISM of galaxies and protogalaxies and are the principal sites of star formation in the universe. There are excellent reasons to adopt this as a working hypothesis. First, the DLA systems have, by definition, $N_{\rm HI} \gtrsim 10^{20}$ cm^{-2}, and this, at least at low redshifts, coincides roughly with a threshold for the onset of star formation (Kennicutt 1989). Second, the DLA systems contain at least 80% of the HI in the universe and appear to be mostly neutral (Lanzetta et al. 1995). The other absorption-line systems, those with $N_{\rm HI} \lesssim 10^{20}$ cm^{-2}, probably contain more gas in total than the DLA systems, but this must be diffuse and mostly ionized. In the following, we regard non-DLA systems as belonging to the IGM, even though some of them might actually be associated with the outer, tenuous parts of galaxies. This distinction—between the mostly-neutral ISM, where stars form, and the mostly-ionized IGM, where they do not—is clearly valid at the present epoch. Thus, it seems appropriate to refer to the DLA systems as DLA galaxies. The precise nature of these objects—whether they large or small, disk or spheroid—remains to be determined, probably by direct imaging. However, as we have already emphasized, the global properties derived from equation (1) are not affected by this uncertainty.

The sample of known DLA galaxies now includes about 80 objects (Wolfe et al. 1995). They are distributed over a wide range in redshift, $0 \lesssim z \lesssim 4$, although, as a consequence of selection effects, most of them are confined to the narrower range $2 \lesssim z \lesssim 3$. From observations of DLA galaxies in various subsets of this sample and comparisons with present-day galaxies, the following trends have emerged. The mean comoving density of HI decreases by almost an order of magnitude, from $\Omega_{\rm HI} \approx (1-2)\times 10^{-3}h^{-1}$ at $z \approx 3$ to $\Omega_{\rm HI} \approx 2 \times 10^{-4}h^{-1}$ at $z = 0$, with $h \equiv H_0/(100$ km s^{-1} Mpc$^{-1})$ (Lanzetta et al. 1995; Wolfe et al. 1995; Storrie-Lombardi et al. 1996). The mean metallicity increases by about an order of magnitude, from $Z \approx 0.1 Z_\odot$ or slightly less at $z \approx 2$ to $Z \approx Z_\odot$ at $z = 0$ (Pettini et al. 1994, 1997b; Lu et al. 1996). The mean dust-to-gas ratio increases by a similar factor, while the mean dust-to-metals ratio remains roughly constant at about the present value in the local ISM (Pei et al. 1991; Kulkarni et al. 1997; Pettini et al. 1997a; Welty et al. 1997; Vladilo 1998). The abundances of H_2 and CO generally appear to be lower at $z \gtrsim 2$ than at $z = 0$ (Levshakov et al. 1992; Ge & Bechtold 1997). Unfortunately, the estimates of $\Omega_{\rm HI}$, Z, and D/G are dominated by relatively few systems—those with the highest values of $N_{\rm HI}$, N_m, and N_d. As a result, they are less certain than is sometimes appreciated.

3. Cosmic Chemical Evolution

The global properties defined above are governed by a set of coupled equations, which are sometimes referred to as the equations of cosmic chemical evolution. In the approximation of instantaneous recycling (and $Z \ll 1$), they take the form

$$\frac{d}{dt}(\Omega_g + \Omega_s) = \dot{\Omega}_f, \tag{2}$$

$$\frac{d}{dt}(Z\Omega_g) + (Z - y)\frac{d}{dt}\Omega_s = Z_f\dot{\Omega}_f, \tag{3}$$

where y is the IMF-averaged yield. These are the cosmological analogs of the usual equations for the chemical evolution of individual galaxies (Pagel 1997). They are exact in the limit that all galaxies evolve in the same way but otherwise must be regarded as approximations. The "source" terms on the right-hand sides of the equations allow for the exchange of material between the ISM of galaxies and the IGM; they represent the inflow or outflow of gas with metallicity Z_f at a rate $\dot{\Omega}_f$. To illustrate a range of possibilities, we consider three types of evolution: a closed-box model ($\dot{\Omega}_f = 0$), a model with inflow of metal-free gas ($\dot{\Omega}_f = +\nu\dot{\Omega}_s$, $Z_f = 0$), and a model with outflow of metal-enriched gas ($\dot{\Omega}_f = -\nu\dot{\Omega}_s$, $Z_f = Z$). Our inflow and outflow models are the cosmological analogs of the standard models of chemical evolution in the disk and spheroid components of the Milky Way (Larson 1972; Hartwick 1976). We fix the yield y in each model by requiring $Z = Z_\odot$ at $z = 0$. Then the only adjustable parameters are the "initial" comoving density of gas in galaxies $\Omega_{g\infty}$ (in practice, the value of Ω_g at $z \gtrsim 4$) and the relative inflow or outflow rate ν.

To complete the specification of the models, we make two other approximations, both motivated by the observations summarized in the previous section: (1) We neglect any ionized or molecular gas in the ISM of galaxies and set $\Omega_g = 1.3\Omega_{\rm HI}$ (to account for He). (2) We assume that just over half of the metals in the ISM are depleted onto dust grains and set $D/G = 0.6Z$. Our models are designed to reproduce (as input) the observed decrease in the mean comoving density of HI between $z \approx 3$ and $z = 0$. The main complication here is that the observed values of $\Omega_{\rm HI}$ tend to underestimate the true values as a consequence of the obscuration of quasars by dust in foreground galaxies (Fall & Pei 1993). We make a self-consistent correction for this bias in the models by linking the obscuration of quasars to the chemical enrichment of galaxies. It is worth noting that, while this correction has a substantial effect on $\Omega_{\rm HI}$, especially at $z \sim 1$, it does not entail large numbers of "missing" quasars [only ~20% at $z = 2$ and ~40% at $z = 4$, consistent with observational constraints (Shaver et al. 1996)]. We neglect another potential bias—that caused by gravitational lensing—because it appears not to be important in the existing samples of DLA galaxies (Le Brun et al. 1997; Perna et al. 1997; Smette et al. 1997). Our models reproduce (as output) the observed increase in the mean metallicity between $z \approx 2$ and $z = 0$ without any fine tuning of the parameters $\Omega_{g\infty}$

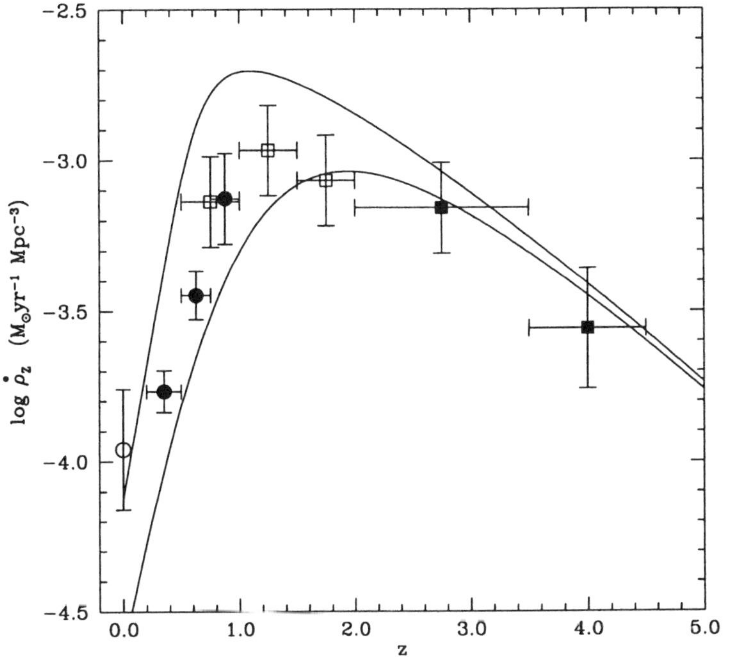

Figure 1. Comoving rate of metal production $\dot{\rho}_z$ as a function of redshift z (for $h = 0.5$, $q_0 = 0.5$, and $\Lambda = 0$). The upper curve represents the inflow model and the lower curve represents the closed-box and outflow models, with $\Omega_{g\infty} = 4 \times 10^{-3} h^{-1}$ and $\nu = 0.5$ [the standard parameters of Pei & Fall (1995)]. The data points represent global Hα and UV emissivities (Gallego et al. 1995; Lilly et al. 1996; Madau et al. 1996, 1998; Connolly et al. 1997).

and ν. The reason for this is that most of the star formation and hence most of the metal production occur at $z \lesssim 2$.

Figure 1 shows the evolution of the comoving rate of metal production in our models, given by $\dot{\rho}_z = y(3H_0^2/8\pi G)\dot{\Omega}_s$. The predicted rates have maxima at $1 \lesssim z \lesssim 2$ and decline rapidly at lower redshifts (Pei & Fall 1995). Figure 1 also shows emission-based estimates of $\dot{\rho}_z$ from several recent surveys, including the Canada-France Redshift Survey and the Hubble Deep Field (Gallego et al. 1995; Lilly et al. 1996; Madau et al. 1996, 1998; Connolly et al. 1997). These were derived from rest-frame Hα and UV emissivities and the approximate proportionality between UV emission and metal production in stellar populations. The main systematic uncertainties in $\dot{\rho}_z$ stem from corrections for incompleteness at low luminosities (which have been included) and corrections for absorption by dust (which have not been included); as a result, the true uncertainties are probably larger than indicated by the error bars in Figure 1. Evidently, the predicted and observed rates are in broad qualitative, and even some quantitative, agreement (within factors of about two). This is remarkable because our models were constructed only with absorption-line observations in mind, before the emission-based estimates of $\dot{\rho}_z$ were available. We have also combined our chemical evolution models with stellar population synthesis models to compute directly the mean comoving emissivity $\mathcal{E}_\nu$ and, by an integration over redshift, the corresponding mean intensity of background radiation J_ν (Fall et al. 1996). These calculations include a self-consistent treatment of the absorption and reradiation of starlight by dust. The predicted far-IR/sub-mm background is consistent with recent results from the DIRBE and FIRAS experiments on *COBE* (Hauser 1996; Puget et al. 1996).

References

Connolly, A.J., Szalay, A.S., Dickinson, M., SubbaRao, M.U., & Brunner, R.J. 1997, ApJ, 486, L11

Fall, S.M., Charlot, S., & Pei, Y.C. 1996, ApJ, 464, L43

Fall, S.M., & Pei, Y.C. 1993, ApJ, 402, 479

Gallego, J., Zamorano, J., Aragón-Salamanca, A., & Rego, M. 1995, ApJ, 455, L1

Ge, J., & Bechtold, J. 1997, ApJ, 477, L73

Hartwick, F.D.A. 1976, ApJ, 209, 418

Hauser, M.G. 1996, in Unveiling the Cosmic Infrared Background, ed. E. Dwek (Woodbury: Am. Inst. Phys.), 11

Kennicutt, R.C. 1989, ApJ, 344, 685

Kulkarni, V.P., Fall, S.M., & Truran, J.W. 1997, ApJ, 484, L7

Lanzetta, K.M., Wolfe, A.M., & Turnshek, D.A. 1995, ApJ, 440, 435

Larson, R.B. 1972, Nature Phys. Sci., 236, 7

Le Brun, V., Bergeron, J., Boissé, P., & Deharveng, J.M. 1997, A&A, 321, 733

Levshakov, S.A., Chaffee, F.H., Foltz, C.B., & Black, J.H. 1992, A&A, 262, 385

Lilly, S.J., Le Fèvre, O., Hammer, F., & Crampton, D. 1996, ApJ, 460, L1

Lu, L., Sargent, W.L.W., Barlow, T.A., Churchill, C.W., & Vogt, S.S. 1996, ApJS, 107, 475

Madau, P., Ferguson, H.C., Dickinson, M.E., Giavalisco, M., Steidel, C.C., & Fruchter, A. 1996, MNRAS, 283, 1388

Madau, P., Pozzetti, L., & Dickinson, M. 1998, ApJ, in press (astro-ph/9708220)

Pagel, B.E.J. 1997, Nucleosynthesis and Chemical Evolution of Galaxies (Cambridge: Cambridge University Press)

Pei, Y.C., & Fall, S.M. 1995, ApJ, 454, 69

Pei, Y.C., Fall, S.M., & Bechtold, J. 1991, ApJ, 378, 6

Perna, R., Loeb, A., & Bartelmann, M. 1997, ApJ, 488, 550

Pettini, M., King, D.L., Smith, L.J., & Hunstead, R.W. 1997a, ApJ, 478, 536

Pettini, M., Smith, L.J., Hunstead, R.W., & King, D.L. 1994, ApJ, 426, 79

Pettini, M., Smith, L.J., King, D.L., & Hunstead, R.W. 1997b, ApJ, 486, 665

Puget, J.-L., Abergel, A., Bernard, J.-P., Boulanger, F., Burton, W.B., Désert, F.-X., & Hartmann, D. 1996, A&A, 308, L5

Shaver, P.A., Wall, J.V., Kellerman, K.I., Jackson, C.A., & Hawkins, M.R.S. 1996, Nature, 384, 439

Smette, A., Claeskens, J.-F., & Surdej, J. 1997, New Astronomy, 2, 53

Storrie-Lombardi, L.J., McMahon, R.G., & Irwin, M.J. 1996, MNRAS, 283, L79
Vladilo, G. 1998, ApJ, 493, 583
Welty, D.E., Lauroesch, J.T., Blades, J.C., Hobbs, L.M., & York, D.G. 1997, ApJ, 489, 672
Wolfe, A.M., Lanzetta, K.M., Foltz, C.B., & Chaffee, F.H. 1995, ApJ, 454, 698

CONFERENCE SUMMARY*

B.E.J. PAGEL
NORDITA, Blegdamsvej 17, Dk-2100 Copenhagen Ø, Denmark

Ocober 1997

Key words: nucleosynthesis, stellar populations, abundance patterns, galaxy evolution, hot gas, stellar yields, globular clusters, light elements

Introduction

As we have heard in the course of this Symposium, there have been major advances in the last few years, especially on the observational side, thanks to HST, Keck, ASCA and also the humble 4m telescopes, among others. As Jim Truran pointed out in his opening review, we can now begin to connect local observations with phenomena at high red-shift, giving a time axis for cosmic, as opposed to merely local, chemical evolution.

Primordial Nucleosynthesis

Dave Schramm brought us up to date on the current situation with primordial abundances. The best baryometer is $(D/H)_P$ (despite some reservations expressed by Jean Audouze) and the most widely accepted current value is 3.2×10^{-5}, by Scott Burles and David Tytler from two 'clean' Lyman-limit systems, although a much higher value is reported for one similarly clean object with $z \simeq 0.7$ observed by K. Lanzetta and his colleagues with HST. Jeff Linsky showed some beautiful Ly-α observations of nearby stars showing the influence of the dynamics of the local interstellar medium. As discussed by Nikos Prantzos, his current value 10^5 D/H $= 1.5 \pm 0.1$ is compatible with Burles & Tytler's primordial value on the basis of reasonable galactic chemical evolution models, but the evolution of ^{3}He is not understood.

The observation of ^{6}Li in the low-metallicity star HD 84937 suggests that the ^{7}Li plateau does not underestimate primordial lithium by more than a factor 2 when uncertainties in the effective temperature scale are taken into account. It would be reassuring to have a similar plateau for D/H. Given Tytler & Burles's upper limit of 4×10^{-5}, the primordial helium abundance Y_P has to exceed 0.243, close to the upper limit estimated by Pagel *et al.* (1992), but in better agreement with the more recent determination by Izotov, Thuan and Lipovetsky (1997). The baryon density parameter $\Omega_{b0}h_{50}^2$ is

IAU Symposium no. 187, *Cosmic Chemical Evolution*, Kyoto August 1997

K. Nomoto and J.W. Truran (eds.), Cosmic Chemical Evolution, 209–218.

Table I
Inventory of baryonic matter

	gm cm^{-3}	$M_\odot$ Mpc^{-3}	Ω_0
$\rho_{\rm crit}$	$4.7 \times 10^{-30}\, h_{50}^2$	$7.7 \times 10^{10}\, h_{50}^2$	
$\rho_b(\frac{\rm D}{\rm H} = 3 \times 10^{-5})$	3.8×10^{-31}	6.2×10^9	$.08\, h_{50}^{-2}$
E/S0$_{\rm lum}$	$7.4 \times 10^{-33}\, h_{50}^2$	$1.2 \times 10^8\, h_{50}^2$	.0015
S$_{\rm lum}$	$3.5 \times 10^{-33}\, h_{50}^2$	$5.7 \times 10^7\, h_{50}^2$	.0007
Cluster gas	$4.4 \times 10^{-33}\, h_{50}^{0.7}$	$7.4 \times 10^7\, h_{50}^{0.7}$	$.001\, h_{50}^{-1.3}$
Group gas	$3.2 \times 10^{-33}\, h_{50}^{0.7}$	$5.2 \times 10^7\, h_{50}^{0.7}$	$.0006\, h_{50}^{-1.3}$
Total lum + gas	$1.8 \times 10^{-32}\, h_{50}^{1.6}$	$3.0 \times 10^8\, h_{50}^{1.6}$	$.0037\, h_{50}^{-0.4}$
$\rho_{\rm Z}(\langle Z\rangle = .02)$		$6.0 \times 10^6\, h_{50}^{1.6}$	$7.4 \times 10^{-5}\, h_{50}^{-0.4}$
Damped Ly-α			$.004\, h_{50}^{-1}$
Gals + DM halos ($\frac{M}{L} = 150\, h_{50}$)			.25
All matter ($f_b = .009 + .14\, h_{50}^{-1.5}$)			$.54\, h_{50}^{-0.7}$
Time-averaged $\dot\rho_*$		$1.5 \times 10^{-2}\, h_{50}^3$ yr^{-1}	
Time-averaged $\dot\rho_{\rm Z}$ ($\langle Z\rangle = .02$)		$4.5 \times 10^{-4}\, h_{50}^{2.6}$ yr^{-1}	

thus between 0.06 and 0.10, requiring substantial amounts of both baryonic and non-baryonic dark matter (see Table I).

Other routes to Ω_{b0} are the baryon fraction in clusters of galaxies, consistent with $\Omega_{b0} = 0.05$, $\Omega_{\rm M0} = 0.25$ (no baryonic catastrophe, but $\Omega_{\rm M} < 1$ if distributed like galaxies; cf. Bahcall *et al.* 1995), large-scale streaming motions and the spectrum of microwave background fluctuations.

I present some numbers relevant to BBNS limits and cosmic chemical evolution in Table I.

The table shows that Ω_{b0} now seems to exceed the contribution from visible stars and gas (Persic & Salucci 1992) by a factor of about $20h_{50}^{-1.6}$ — a vast amount of baryonic dark matter that may be accounted for by ionized gas associated with the Ly-α forest clouds (e.g. Weinberg *et al.* 1997), but has no well-known counterpart today (intergalactic gas? Machos? LSB galaxies?). The total Ω_{M0} from galaxy halos and cluster dynamics is about 0.2 to 0.3 according to Bahcall *et al.* (1995) and from the baryon fraction in clusters of galaxies according to White *et al.* (1993) it is $\leq 0.54h_{50}^{-0.7}$. The other entries in the table will be commented on later.

Population III

Population III was postulated 30 years ago by Doroshkevich, Zeldovich and Novikov to explain the re-ionization of the intergalactic medium and quite widely discussed for some years, but then widely neglected. Jerry Ostriker has now brought the idea back in force by showing that pre-galactic stars can indeed be formed at red-shifts up to 15 and produce a wide range of initial metallicities in different places according to density and other factors.

Observed galaxies including Lyman forest clouds are found with metallicities down to about 10^{-3} solar (Tytler) whereas in our own Galaxy stars have been found down to 10^{-4}. Are these latter a genuine Population III, possibly associated with the mini-halos suggested by Mesashi Chiba, or can we expect to find Lyman clouds of still lower metallicity at still higher red-shifts than observed to date?

Ostriker mentioned an interesting test: if the most metal-poor stars in our Galaxy are pregalactic, they may be expected to have a dark-matter like density distribution $\sim R^{-2}$ in contrast to the steeper fall-off of the conventional stellar halo.

Galaxy formation and evolution

This is a huge topic, to which one can hardly do justice in a short summary. As was explained to us by Simon White and Richard Ellis, it is currently driven by observations, notably the AAT-LDSS and Canada-France red-shift surveys to $z \simeq 1$ and the discovery of star-forming galaxies up to $z \simeq 4.5$ using the multi-colour drop-out technique, especially in the Hubble Deep Field, with red-shifts confirmed using Keck.

At the lower red-shifts, it appears that regular large galaxies like our own were already established by $z = 1$ at which time they were about 1^{m} brighter than now; ellipticals are consistent with passive evolution although on-going mergers are not excluded. On the other hand, there is a dramatic rise in the number of blue irregular star-forming galaxies with [O II] emission.

At higher red-shifts, Madau *et al.* (1996) have estimated the luminosity and hence the star-formation-rate density as a function of z from the HDF, and there are signs of a decrease beyond $z = 2$, corresponding to a peak near $z = 1$ as predicted by Pei & Fall on the basis of the decline in $\Omega_{\mathrm{H\,I}}$ with decreasing z. According to the models and according to the analysis by Storrie-Lombardi *et al.* (1996), $\Omega_{\mathrm{H\,I}}$ is more or less constant (or at least not increasing with z) for $z > 2$, so disk galaxies would have been forming few stars earlier than that. Madau *et al.* found fair agreement with an average 'fiducial' star formation rate giving a solar average metallicity to visible matter at the present time (see $\rho_Z(\langle Z \rangle = .02)$, time-averaged $\dot{\rho}_*$ and time-averaged $\dot{\rho}_Z(\langle Z \rangle = .02)$ in Table I), but there are problems with incompleteness and absorption so that all the star formation rate estimates are probably too low (e.g. Pettini *et al.* 1997). That this is so is also suggested by the ISO survey of HDF galaxies by Rowan-Robinson *et al.* (1997), who find larger star formation rates by a factor of about 3 and suggest that the extra metals are hidden in baryonic dark matter. Thus we may have a dark metal problem as well as a dark matter problem!

The evolving galaxies at high red-shift tend to be compact and may be merging to form more conventional galaxies as was described very nicely by Simon White and Guinevere Kauffmann, but this is still uncertain.

Evolution in our own Galaxy has been milder, as shown by Nikos Prantzos; this is entirely consistent with the evolution of spirals described by Richard Ellis. Elliptical galaxies can arise from mergers of disk-like systems, as Guinevere described, but these were not the spirals that we see today. Another type of merger is on-going, at least in the Milky Way halo, as described by S. Majewski, and leads to an uneven distribution of the stars in phase space.

After all this, it is not clear to me how the classical ELS model of formation of our Galaxy, still supported by the properties of the inner halo and thick disk if not by those of the outer halo, ties in with current models of hierarchical clustering.

Abundance patterns

These have been very nicely reviewed for us by Chris Sneden for the halo and metal-poor thick disk, by Mike Rich for the Bulge and by Limin Lu for the damped Lyman-α systems, which (as Jim Truran pointed out) can now provide a time-scale for metal enrichment, although (as shown by Lanzetta) the age-metallicity relation derived for the latter demands caution because the geometry favours high impact parameters (cf. Phillipps and Edmunds 1996). The apparent lack of increase in mean metallicity from $z = 2$ to $z = 1$ could be due to this effect.

The basic pattern of α-element enrichment at low metallicities, first noticed by Aller & Greenstein nearly 40 years ago, persists among Galactic stars and apparently also shows up in Lyman forest clouds (Si/C observed by Tytler and by Boksenberg); whether it does so in damped Lyman-α systems is controversial. Don York explained that corrections from observed to total abundance are tricky owing to the presence of a multi-phase medium with different ionization and dust properties.

The explanation of the α/Fe effect in terms of a time delay of the order of 1 Gyr or more for SN Ia which provide about 60 per cent of the iron in the Solar System still seems to hold good (although possibly challenged by Takuji Tsujimoto's report of a high Fe/Mg ratio in a quasar), but among Galactic halo stars there are exceptions as noted recently by Nissen & Schuster (1997) and by King (1997) and as can also be retrieved from earlier data. The exceptional stars were presumably captured from merging dwarf galaxies for which Majewski provided ample evidence; in these systems, e.g. the LMC as reported by Beatriz Barbuy, the effect is mitigated by slower star formation and possibly effects of bursts separated by quiescent intervals (Gilmore & Wyse 1991). The main difference observed between damped Lyman-α systems on the one hand and both Galactic halo stars and dwarf irregular galaxies on the other is the low N/Si ratio, accompanied by a large scatter, in the DLA systems. This favours a scenario in which the stars in the DLA systems are very young and primary nitrogen comes from intermediate-mass stars with a significant time delay, rather than from primary nitrogen production by mixing processes in massive Population III stars.

The behaviour of heavy metals in Galactic halo stars is intriguing. For $[Fe/H] \leq -3$, there is a strange pattern among iron-group elements (which may or may not be explicable by supernova nucleosynthesis models) and a large scatter in the abundances of neutron-capture elements relative to iron. However, the neutron capture elements themselves often show an excellent match to the r-process contributions in the Solar System, notably in the case of CS 22892–092, which is a sort of Rosetta stone for the r-process and has been used as a thorium chronometer.

Above $[Fe/H] = -3$, the r-process continues to dominate for a while (as Jim Truran predicted 15 years ago), notably for barium in HD 140283 and 122563, and this is now confirmed by hyperfine structure broadening as shown in the very nice new work by Walter Gacquer. By the time [Fe/H] reaches -2, however, the s-process takes over for barium and other conventional s-process elements, which is not exactly what is expected from current models of low-mass AGB stars. Pagel & Tautvaišiene (1997) have modelled the Galactic chemical evolution of these elements assuming two distinct s-processes with different time delays; only the second one, with a time-delay exceeding that of SN Ia, corresponds to the classical AGB model.

Among Galactic disk stars, as described by Bengt Edvardsson, the old discrepancy between abundance gradients derived from B stars and from other sources (H II regions, F stars) seems to be going away (Smartt and Rolleston 1997). The dispersion in metallicity at a given age is ± 0.1 to ± 0.15 dex, a bit less than some of the values in the classic Edvardsson *et al.* (1993) paper where they already pointed out that the scatter was slightly enhanced by selection effects. The solar oxygen abundance is about 1.5 times higher than the concordant determinations in the Orion nebula (Esteban *et al.* 1997), in B-type stars (e.g. Gies and Lambert 1992) and in the diffuse interstellar medium (Meyer, Jura and Cardelli 1997). The scatter could arise from orbital diffusion or gas inflow episodes; similarity of scatter between Fe and α-elements rules out a dominant contribution from self-enrichment in stellar associations. Recent results by B. Nordström *et al.* give a mild increase in C/Fe with decreasing Fe/H, indicating a major source of carbon in relatively massive stars.

Coming to the Bulge, as described by Mike Rich, things are quite confusing. Because star formation there was presumably more rapid than in our neighbourhood, we expect the α-rich effect to persist to higher metallicities. Mg and Ti seem to behave as expected, but Ca and Si do not, while oxygen is too difficult to measure. The abundance distribution function in terms of Mg roughly fits a Simple model with a large effective yield, which could arise from a dissipative collapse of the halo-bulge system setting up an abundance gradient. There is an intriguing connection with a bunch of super metal-rich field stars discovered by Michel Grenon and the SMR old open cluster (NGC 6791) with a blue horizontal branch reported by Ruth Peterson; this can now be used to calibrate SMR giant luminosities in external galaxies.

Hot gas in ellipticals and in groups and clusters of galaxies

The ASCA satellite has led to major developments in the study of hot X-ray gas. Possibly the most dramatic is the discovery reported by Mikato Hattori of a dark, massive gravitationally lensing cluster emitting X-rays with the usual iron line at $z \simeq 1$ and little optical light from a sole cD galaxy. The baryon fraction f_b is only half of the value quoted in Table I. Where did the iron come from if there are so few stars? There may also be a massive but cold cluster (no X-rays) at $z = 0.75$.

Gas phase abundances in more conventional clusters and individual elliptical galaxies were discussed from the observational point of view by Takaya Ohashi and M. Loewenstein and from the theoretical point of view in a beautiful presentation by Ms Yuhri Ishimaru. In some cases, the distribution of iron with radius tends to match that of the galaxies, which is steeper than that of the gas, and in general the Fe/O or Fe/α pattern is more or less solar, implying some contribution from SN Ia. In any case, while the iron abundance in the hot gas of rich clusters is around 1/3 solar, the total mass of iron, relative to stellar luminosity and presumably mass, is very large, indicating larger stellar yields than we think we have in the solar neighbourhood, although the mean metallicity of the stars themselves is not above solar. Ishimaru and Arimoto have developed theoretical models involving galactic winds and group winds, assuming high stellar yields (top-heavy IMF), with only the rich clusters managing to retain most of their gas.

Stellar yields

This leads on to the stellar yields themselves. Dave Arnett discussed SN II, including those associated with Population III, and emphasized the crucial role played by mixing in shell O-burning zones, which can now be treated in direct hydrodynamical simulations. The old onion-skin picture is challenged by the SN 1987A light curve and by isotopic anomalies (^{26}Al, ^{44}Ti) in presolar grains. Yields of primary elements like oxygen are not greatly affected, but those of secondary elements like nitrogen are very sensitive to mixing details. Dave was kind enough to quote me as having called for a primary nitrogen source for low-metallicity objects, but in view of the DLA behaviour I would rather favour intermediate-mass stars as the source.

Ken Nomoto and Xiangdong Li discussed SN Ia, which are now thought to come in two versions, both favouring a single-degenerate model with carbon deflagration at the Chandrasekhar limiting mass. The new type involves rapid mass transfer in systems identified with ultra-soft X-ray binaries.

AGB stars were discussed by Nami Mowlavi with special reference to ^{3}He, ^{26}Al, Tc, F and ^{7}Li, among others. Here again, extra mixing is required, in particular to produce the third dredge-up at low enough luminosities, to generate neutrons by $^{13}\mathrm{C}(\alpha, n)$ and to account for the destruction of ^{18}O. AGB stars may account for ^{3}He destruction and formation of ^{19}F and ^{26}Al, although Wolf-Rayet stars could be a better source for fluorine.

Brad Gibson gave an enlightening account of the wide range in stellar yields calculated by different authors, confirming my prejudice that one can to a large extent just as well take the yields *ad hoc* by fitting abundances with galactic chemical evolution models (cf. Pagel and Tautvaišiene 1995). M. Samland presented a highly sophisticated model in which he did just that.

New galactic chemical evolution models

Apart from the empirical determination of yields by Samland and the work on elliptical galaxies and hot gas, we had a fine presentation of the chemical evolution of the Galaxy by Nikos Prantzos and we heard briefly from Beatriz Barbuy about the LMC. Cosmic chemical evolution models are the Holy Grail as far as this meeting is concerned, tying together the information that we have, ranging from local galaxies to the universe at high red-shift. The two we have heard about are from Uta Fritze and Mike Fall who look at the question from quite different viewpoints. Uta uses basically the stellar population synthesis approach, tying the scattered metallicities in DLA and C IV systems respectively to disks and halos in different morphological types of galaxies seen today, whereas Mike Fall deals with a global average, tying star formation to the declining density of H I. The effects of varying ionization, and the status of elliptical galaxies, are not clear in these models, although such complications are given some recognition in Mike's inflow and outflow parameters.

Globular clusters

Globular clusters were discussed from the theoretical point of view by Doug Lin and from the observational point of view by Jean Brodie. Doug reached the interesting conclusion that globular clusters are already born with mass segregation, the most massive stars lying at the centre. It was not clear to me whether the predicted dependence of minimum stellar mass on metallicity could have some observational consequences.

Jean Brodie presented observations of the abundance distribution functions (ADFs) in extragalactic globular cluster systems (NGC 1275 and 1399) designed to test the hypothesis that globular clusters in elliptical galaxies are formed in mergers. Cluster ADFs in ellipticals are bimodal, with the more metal-rich clusters resembling the field stars; with other evidence, this suggests a two-phase model rather than a merger model, although Richard Larson contributed a percipient suggestion that the bimodal distribution could come from mergers with a galaxy like our own. Three clusters in NGC 1275 have A-type spectra and seem to demand a peculiar truncated IMF; these are not expected to survive long enough to develop into conventional old globular clusters, although here again there are caveats.

Light elements

Abundances of ^{6}Li, ^{9}Be and ^{10}B in the Solar System are readily explained by cosmic-ray spallation on CNO nuclei in the interstellar medium, but there are problems with the $^{10}B/^{11}B$ ratio and with the constant abundances of Be and B relative to iron or oxygen in low-metallicity stars. The solution currently proposed by Cassé, Vangioni-Flam, Ramaty *et al.* is based on energetic C and O nuclei in supernova ejecta impacting protons in the local interstellar medium.

There is still a problem, in that the SN ejecta need to be hydrogen-poor, coming from WC star progenitors, and Nikos Prantzos pointed out that these become rare in metal-poor systems, so the problem is not completely solved after all. Francesca Primas drew attention to another mystery: why do some stars have normal lithium but reduced beryllium and boron?

Conclusion

We have heard many interesting things at this very successful symposium. No problem has been solved in a conclusive sense, but there has been dramatic progress in the past three years and we now have some insights into the deep universe from the chemical evolution point of view. No doubt there will be yet another sea-change in knowledge and viewpoints by the time of the next IAU.

Finally, let us express our thanks to Jim Truran for the excellent work he has done to make this symposium a success and to our colleagues on the LOC for outstanding efficiency and hospitality which has made this IAU assembly such an enjoyable occasion.

References

Bahcall, N., Lubin, L.M. and Dorman, V. 1995, *Ap. J. Lett.*, **447**, L81
Edvardsson, B., Andersen, J., Gustafsson, B., Lambert, D.L. and Tomkin, J. 1993, *A & A*, **275**, 101
Esteban, C., Peimbert, M., Torres-Peimbert, S. and Escalante, V. 1997, *MNRAS*, submitted
Gies, D.R. and Lambert, D.L. 1992, *Ap. J.*, **387**, 673
Gilmore, G. & Wyse, R.F.G. 1991, *Ap. J. Lett.*, **367**, L55
Izotov, Y.I., Thuan, T.X. and Lipovetsky, V.A. 1997, *Ap. J. Suppl.* **108**, 1
King, J.R. 1997, *Astr. J.*, **113**, 2302
Madau, P., Ferguson, H.C., Dickinson, M.E., Giavalisco, M., Steidel, C.C. and Fruchter, A. 1996, *MNRAS*, **283**, 1388
Meyer, D.M., Jura, M. and Cardelli, J.A. 1997, *Ap. J.*, in press
Nissen, P.E. & Schuster, W.J. 1997, *A & A*, in press
Pagel, B.E.J., Simonson, E.A., Terlevich, R.J. and Edmunds, M.G. 1992, *MNRAS*, **255**, 325
Pagel, B.E.J. and Tautvaišiene, G. 1995, *MNRAS* **276**, 505
Pagel, B.E.J. & Tautvaišiene 1997, *MNRAS*, **288**, 108
Persic, M. and Salucci, P. 1992, *MNRAS*, **258**, 14P
Pettini, M., Steidel, C.C., Adelberger, K.L., Kellogg, M., Dickinson, M. and Giavalisco, M. 1997, in *Origins*, ASP Conf. Series
Phillpps, S. and Edmunds, M.G. 1996, *MNRAS*, **281**, 362
Rowan-Robinson, M., Mann, R.G., Oliver, S.J. *et al.* 1997, *MNRAS*, **289**, 490
Smartt, S.J. and Rolleston, W.R.J 1997, *Ap. J. Lett.*, **481**, L47
Storrie-Lombardi, L.J., MacMahon, R.G. and Irwin, M.J. 1996, *MNRAS*, **283**, L79

Weinberg, D.H., Miralda-Escudé, J., Hernquist, L. and Katz, N. 1997, preprint astro-ph/97 01012
White, S.D.M., Navarro, J.F., Evrard, A.E. and Frenk, C.S. 1993, *Nature*, **366**, 429

NITROGEN ABUNDANCE IN OXYGEN-RICH GIANTS THROUGH K TO LATE M

W. AOKI, T. TSUJI
Institute of Astronomy, The University of Tokyo
Osawa, Mitaka, Tokyo, 181 Japan

Nitrogen abundances in red giants are important to understand the contribution of low-mass and intermediate-mass stars to the chemical enrichment of the Galaxy. Nitrogen abundances in 23 oxygen-rich cool evolved stars are derived by the analyses of high resolution infrared spectra including CN lines (red systems) in the K- and H-window regions as well as NH lines in the L-window region (Aoki W. & Tsuji T. 1997, A&A, in press). Our sample includes 2 K giants, 20 M giants and 1 S-type star.

The result shows that the nitrogen abundances in late M giants are larger than those in early M giants while decrease of the carbon abundance was found in late M giants by our previous work (Tsuji T., 1991, A&A 245, 203). The average abundances in early and late M giants are shown in the table (log $A_{\rm H}$=12). These variations of abundances can not be explained by the first dredge-up model but require additional processing by the CN cycle and mixing after the first dredge-up. However, there is no obvious evidence of other processes such as the 3α-process and subsequent hot bottom burning in our program stars. Such variation of the abundances is not well understood by the present evolutionary models of low-mass and intermediate-mass stars.

spectral type	number of objects	log $A_{\rm C}$	log $A_{\rm N}$
early M giants (K5-M3.5)	11	8.37	8.11
late M giants (M4-)	8	8.10	8.45

Nitrogen would be produced in massive stars which evolve to supernovae as well as in low-mass and intermediate-mass stars by the CNO cycle, but their contributions to the chemical evolution of the Galaxy are not resolved quantitatively. Our result which shows high nitrogen abundance in late M giants is one of the bases to understand the origin of nitrogen.

ASYMMETRICAL DELAYED DETONATION FROM A 3D HYDROSIMULATION OF A WHITE DWARF EXPLOSION

E. BRAVO, AND D. GARCíA-SENZ
Dpt. Física i Enginyeria Nuclear, UPC, Barcelona, Spain, and Institut d'Estudis Espacials de Catalunya, Barcelona, Spain

The White Dwarf Delayed Detonation model for SNIa has been investigated using an SPH code. The relevant features and results are:

- The code is able to handle deflagrations as well as detonations.

- The code incorporates a nuclear network of 9 nuclei, from ^{1}H to ^{56}Ni. The final nucleosynthesis comes from a post-processing calculation.

- During the deflagration phase the combustion front is increasingly deformed, but the effective velocity of the flame is too low to unbind the whole white dwarf. When the density at the flame reaches $\sim 5 \times 10^7$g/cm^3 the fractal dimension of the flame surface at selected points is greater than 2.5. These points are then artificially incinerated, starting the detonation phase.

- The shock waves launched by the incineration of the high fractal dimension regions are able to induce detonation waves in the fresh fuel only where the environmental conditions are propitious (geometry, density, etc).

- The final composition is rich in C-O (0.76 $M_{\odot}$), whereas the ejected mass of ^{56}Ni (0.26 $M_{\odot}$) is slightly too low for a SNIa. The detailed composition is underabundant in C-O, and has nearly Solar System proportions of ^{32}S, ^{36}Ar, ^{40}Ca, ^{48}Ti, ^{52,53}Cr, ^{55}Mn, ^{56,57}Fe, ^{60,61,62}Ni, and ^{64}Zn, while ^{54}Fe and ^{58}Ni are a factor 4-5 in excess.

- As a result of the asymmetrical propagation of the detonations, the typical distribution of abundances in the velocity space obtained in 1D models is not reproduced here. Thus, Ni, Si, and S are more abundant than C and O in the high velocity material, from $\sim 13,000$ to $27,000$ km/s. Between $3,000$ and $\sim 13,000$ km/s C, O, and Ni are the dominant nuclei, while at lower velocities the composition is rich in Fe.

As a conclusion, the feasibility of the computation of a Delayed Detonation in 3D has been shown. Our results show that the multidimensional properties of the combustion front must be carefully taken into account, which can not be done with a 1D code. Obviously, much work remains to be done.

NUMERICAL SIMULATIONS OF THE CHEMICAL EVOLUTION OF GALAXIES USING TREECODE-SPH

D. K. CHURCHES, A. H. NELSON AND M. G. EDMUNDS
Department of Physics and Astronomy, University of Wales, Cardiff
PO Box 913, Cardiff CF2 3YB

Using an SPH-Treecode which has been modified to include star formation and heavy element synthesis we are investigating the chemical evolution of galaxies from within an overall picture of galaxy formation. Our aim is to understand which chemical properties of galaxies can be explained using a simple collapsing model with star formation.

Our initial conditions are a uniform density sphere with Poisson noise in solid body rotation, and a mass of $2.5 \times 10^{12} M_{\odot}$. Nine tenths of this mass is in the form of dark matter, one tenth is gaseous. This collapses down to form a gas disk of radius $\sim$ 12 kpc. The star formation rate is given by a Schmidt Law, $SFR = k\rho_g^n$, where $1 \leq n \leq 2$ and the heavy element synthesis is modelled under the assumption of instantaneous recycling. We study the effects of varying the constant k and index n in the Schmidt Law. This enables us to vary the relative sizes of the star formation and dynamical timescales. To perform the simulations we use 5000 gas particles, 5000 dark matter particles, and we typically form 7000 star particles.

Our initial results show that:

- These initial conditions can lead to the formation of disk galaxies which have surface density profiles which are well described by an exponential distribution. The typical scale length for the disk is $\sim$ 3.5 kpc.
- For a Schmidt Law of index $n = 1$, no abundance gradients are formed, in accordance with the work of Edmunds and Greenhow 1995.
- For a Schmidt Law of index $n = 1.5$, abundance gradients are formed, but they are shallower ($\sim -0.02\, dex\, kpc^{-1}$) than those typically observed in disk galaxies ($\sim -0.08\, dex\, kpc^{-1}$).

Edmunds M. G., Greenhow R. M., MNRAS 1995, 272, 241

CHEMICAL EVOLUTION OF DAMPED LYMAN α SYSTEMS

K. DENDA
National Astronomical Observatory of Japan
2-21-1 Osawa, Mitaka, Tokyo 181, Japan

1. Chemical Evolution Model

I study chemical evolution of damped Lyman α systems (DLAs) as normal disk galaxies and compare the results with the observed redshift - metallicity relation and the relative abundance pattern, [Si/Fe] vs. [Fe/H] (Lu et al. 1996). A critical gas-mass fraction of star formation is $f_{\rm gas}^{\rm crit} = 2.5 \times 10^{-2} \left(\frac{R_{\rm DLA}}{20{\rm kpc}}\right)^2 \left(\frac{10^{11} M_\odot}{M_{\rm total}}\right) \left(\frac{\Sigma_{\rm H}^{\rm crit}}{2 M_\odot {\rm pc}^{-2}}\right)$, $f_{\rm gas}^{\rm crit} = 0,\ 0.1,\ \&\ 0.3$, where $\Sigma_{\rm H}^{\rm crit}$ is the critical surface density of star formation (Kennicutt 1989). Here the dependence both on radius and $v_{\rm rot}$ is neglected, for simplicity. I then assume the formation epoch of DLAs and translate their age into the corresponding redshift in order to compare observations.

2. Results

A chemical evolution model suitable for normal disk galaxies can explain the observed large dispersion in metallicity of DLAs, if galaxy formation continues from $z \simeq 5$ to 3. The threshold of star formation also cases a similar scatter. However, a few DLAs do not follow such picture of evolution. Although relative abundances pattern of the DLA at $z = 1.78$ on the line of sight toward Q1331+17 are consistent with those of stars in our Galaxy and other DLAs, its metallicity is much lower. Such a DLA may be affected by dust depletion or belong to other class of DLAs.

References

Kennicutt, R.C. 1989, ApJ 344, 685
Lu, L., Sargent, W.L.W., Barlow, T.A., Churchill, C.W. & Vogt, S.S. 1996, ApJS 107, 475

ABUNDANCES IN METAL-RICH STARS

Detailed abundance analysis of 47 G and K dwarf stars with [Me/H] > 0.10 dex

S. FELTZING
Royal Greenwich Observatory, Madingley Road, Cambridge CB3 0EZ, United Kingdom

We have derived elemental abundances of O, Na, Mg, Al, Si, Ca, Ti, Cr, Mn, Fe, Co, Ni as well as for a number of s-elements for 47 G and K dwarf, with [Me/H]> 0.1 dex. The selection of stars was based on their kinematics as well as on their $uvby - \beta$-photometry. One sample of stars on rather eccentric orbits traces the chemical evolution interior to the solar orbit and another, on circular orbits, the evolution around the solar orbit. A few Extreme Population I stars were included in the latter sample.

We have performed a standard (LTE) analysis to derive chemical abundances, for details see Feltzing & Gustafsson (1997) and Feltzing (1995). Our data do, in general, fit well into the overall picture of galactic chemical evolution as it is currently understood from empirical studies. They give further constraints on theoretical models of galactic chemical evolution for the region above solar metallicity, where metallicity dependent supernova yields further complicate the picture. Our lack of success in tracing variations in relative abundances, e.g. of [O/Fe], [Na/Fe], [Si/Fe], [Ti/Fe] etc to vary with the stellar velocities, i.e. with the present orbital mean distance from the galactic centre ($R_{\rm m}$), point to a homogeneous evolution in the disk over the radius spanned. We find that the upturn in [Na/Fe] vs [Fe/H], as observed by Edvardsson et al. (1993), is reproduced and reinforced. In contrast to sodium we find no such upturn for aluminium. This is somewhat surprising since sodium and aluminium are thought to be produced in the same environment in the pre-supernova star.

References

Edvardsson B., Andersen J., Gustafsson B., Lambert D.L., Nissen P.E., Tomkin J. 1993a, A&A 275, 101

Feltzing S., 1995, Thesis, "Two studies of galactic chemical evolution", Uppsala University

Feltzing S., Gustafsson, B.,1997, A&AS, accepted, astro-ph 9710315

ABUNDANCE PATTERN OF SUPERGIANTS IN THE FIELD OF THE SMALL MAGELLANIC CLOUD AND IN THE CLUSTER NGC330

V. HILL
Observatoire de Paris-Meudon
CNRS URA 335, F-92195 Meudon Cedex, FRANCE

NGC330 (the brightest SMC young cluster, with an age of ~20-25 Myrs) contains many bright red and blue supergiants. This cluster has arisen special interest, since the earlier abundance determinations were indicating an abundance lower by a factor ~5 than in the SMC field young population. How could a young cluster form with a metallicity lower than that of the surrounding field ? Moreover, in the SMC, the young field population seems to be fairly well mixed (small abundance scatter among H II regions and young stars). Taking advantage of our previous high resolution analysis of a sample of in the SMC field type K supergiants (Paper I and II), we analyse 6 cool supergiants in NGC330, in order to compare in an homogeneous way, the metallicity and abundance ratios of various elements in NGC330 and in the field of the SMC. The main results may be summarised as follows:

- the deficiency of NGC330 is found to be of [Fe/H]=-0.82±0.11 dex (mean and *rms* for 6 stars) whereas the field stars had a [Fe/H] =-0.69±0.10 dex (mean and rms for 6 stars). A slight difference may exist between the field and the cluster stars, but this difference is small.
- the abundance pattern of the ~20 elements studied, is similar to that of the field supergiants, except for a few elements : Oxygen's star to star scatter is significantly larger in the cluster; the very heavy neutron capture process elements ([La/Fe],[Eu/Fe]...) are less enhanced in the cluster.
- lithium is detected in all stars and is low, indicating that the stars underwent dilution (mean value and *rms* for the 6 stars: log $N_{Li}/N_H \simeq -11.4 \pm 0.2$).

References

Hill, V., (1997) *A&A* 324, 435 (Paper I)
Hill, V., Barbuy, B. and Spite, M. (1997) *A&A* 323, 461 (Paper II)

AGE AND METALLICITY DISTRIBUTION OF THE DISK STARS FROM EDGE-ON GALAXIES

A. JUST
Astron. Rechen-Institut
Mönchhofstr. 12–14, D-69120 Heidelberg, Germany

Due to the increasing velocity dispersion $\sigma(t)$ and vertical thickness of the stellar distribution with age t the composition of the disk population in age and metallicity varies with height above the galactic plane. This results in a central steepening of the vertical luminosity profiles and in characteristic colour gradients, which can be observed in edge-on galaxies. For the analysis of the vertical profiles we use a self-gravitating disk composed of isothermal subcomponents according to a SFR(t) with increasing $\sigma(t)$ and decreasing [Fe/H](t) as a function of age t. The broad band luminosities and metal indices are computed with the photometric evolutionary synthesis method (Just, A., Fuchs, B., Wielen, R. 1996, *Astron. Astrophys.* **309**, 715). The vertical colour profiles give a useful tool to derive the age distribution of the disk, but for the metal enrichment additional spectral information is necessary. In Fig. 1 the discriminating power of the Mg_2-index is shown in an example.

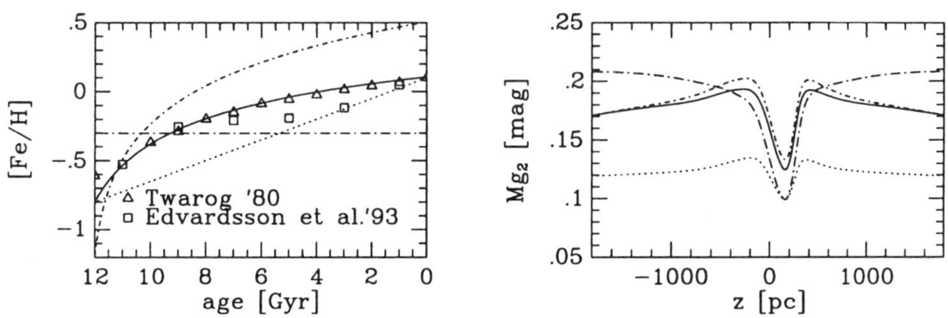

Figure 1. Left: Different metal enrichment functions compared to the observed data of the solar neighbourhood. Right: Vertical profiles of a disk with a SFR(t) and $\sigma(t)$ typical for a late type spiral using the different [Fe/H](t) functions. The inclination is i=89.5^0.

EVOLUTION OF THE COLOUR-MAGNITUDE RELATION OF ELLIPTICAL GALAXIES

TADAYUKI KODAMA[1,2], NOBUO ARIMOTO[2], AMY BARGER[1,3]
AND
ALFONSO ARAGÓN-SALAMANCA[1]
[1] *Institute of Astronomy, University of Cambridge, Madingley Road, Cambridge, CB3 0HA, UK*
[2] *Institute of Astronomy, University of Tokyo, Mitaka, Tokyo 181, Japan*
[3] *Institute for Astronomy, University of Hawaii, 2680 Woodlawn Drive, Honolulu, Hawaii 96822, USA*
e-mail: kodama@ast.cam.ac.uk

This paper presents the quantitative analyses of evolution of the colour-magnitude (C-M) relation of early type galaxies in 9 distant clusters (AC 103 [z=0.31], AC 114 [0.31], AC 118 [0.31], Cl 0024+16 [0.39], Cl 0412-65 [0.51], Cl 0016+16 [0.546], Cl 0054-27 [0.563], GHO 1603+43 [0.895], and 3C 324 [1.206]) at $0.3 < z < 1.2$ based on the evolutionary model of elliptical galaxies (Kodama & Arimoto 1997, A&A, 320, 41). The model is calibrated to reproduce the C-M relaiton of Coma ellipticals at $z = 0$ (Bower, Lucey, Ellis 1992, MNRAS, 254, 589) and gives the evolution of slope and zero-point of the C-M relation as a function of redshift. The slopes of C-M relation can be reproduced by a single model sequence in which mean stellar metallicities of ellipticals are varied as a function of galaxy luminosity and ages are assumed to be equally old. The zero-points of C-M relation of the two most distant clusters (GHO 1603+43, 3C 324) put constraint on the epoch of major star formation in early type galaxies that should be $z > 2 - 4$. The cluster early type galaxies are suggested to be uniformly old and to have universal metallicity sequence as a function of galaxy mass. We find no significant difference of C-M relation among the different clusters. This will put strong constraint on the formation of clusters of galaxies as well, since they must have shared a common formation mechanism to produce the homogeneous C-M relation.

THE OPTICAL PROPERTIES OF CARBON AND GRAPHITE DEPEND ON CRYSTAL STRUCTURE AND TEMPERATURE.

C. KOIKE(1), H. SHIBAI(2), T. TANABE(3), H. SUTO(4) // AND H.SOGAWA(5), C. KAITO(6),Y. ANDO(7), M. FUKADA(1)
(1)Kyoto Pharmaceutical Univ.,Kyoto 607, (2)Nagoya Univ., Nagoya 464-01, (3)Inst.of Astronomy,the Univ.of Tokyo,Tokyo 181, (4)National Astronomical Observ.,Tokyo 181, (5)Kyoto Univ.,Kyoto 606, (6)Ritumeikan Univ.,Shiga 525, (7)Meijyou Univ.,Nagoya 464

We measured the optical properties of various carbon and graphite in wide wavelength from UV to FIR. The correlation between the optical properties at UV and MIR and the degree of crystallization has not been so clear. The correlation between the spectral index at FIR and the degree of crystallization has been clear.The spectral index increases from 0.6 to 1.3 for carbon samples, and from 1.8 to 2.9 for graphite samples. The spectral index of carbon samples increases as the size of microcrystallites increase and crystal structure changes to onion-like. The spectral index of graphite samples increases as the structures of graphite crystal become perfect. These measurements were made at room temperature. Neverthless, the carbonaceous dust grains might be affected by surrounding temperature. So, we measured the spectra at low temperature (He liquid temperature). All samples of amorphous carbon at liquid He temperature show larger spectral index than that at room temperature. The sample of graphite T10, the best crystal structure among samples, shows the same spectral index as at liquid He and room temperature. Another samples of graphite at liquid He show nearly same spectral index as at room temperature. The spectral index in the FIR strongly depends on the degree of crystallization of carbon and graphite samples, and surrounding temperature. We can estimate the degree of crystallization from the spectral index of the absorption of carbon or graphite samples, and furthermore, from the observed index of circumstellar dust around carbon stars. This fact might give the clue to know the thermal history of carbonaceous dusts in the solar system and circumstellar around primitive stars.

IMPROVED THEORETICAL STELLAR LIBRARY FOR POPULATION SYNTHESIS MODELS

T. LEJEUNE AND R. BUSER
Astron. Inst. Univ. Basel, Switzerland

We present a *standard library* for evolutionary and population synthesis studies which provides theoretical stellar energy distributions (SEDs) from UV to IR for the ranges in fundamental parameters, $T_{\rm eff}$ (50,000 K $\sim$ 2000 K), $\log g$ (-1.0 $\sim$ 5.5) and $[M/H]$ (-5.0 $\sim$ +1.0). A correction procedure was applied to the original synthetic spectra yielding color-calibrated flux distributions (Lejeune *et al.* 1997a, 1997b). This stellar library was used to model the integrated light of globular clusters. Comparisons of population models with observed color-magnitude diagrams and integrated spectra of (metal-rich) Bulge globular clusters (Bruzual *et al.* 1997) show that the semi-empirical flux corrections applied to the original synthetic stellar spectra improve the agreement between model predictions and observations (Fig 1). The population models are also in good agreement with observed UBV and CMT_1T_2 colors of galactic and extragalactic clusters (Lejeune, 1997).

Figure 1. Best fit of the integrated spectrum of NGC6528 (dotted line) from population models using (a) the original SEDs (dashed line) and (b) the color-calibrated SEDs (solid line).

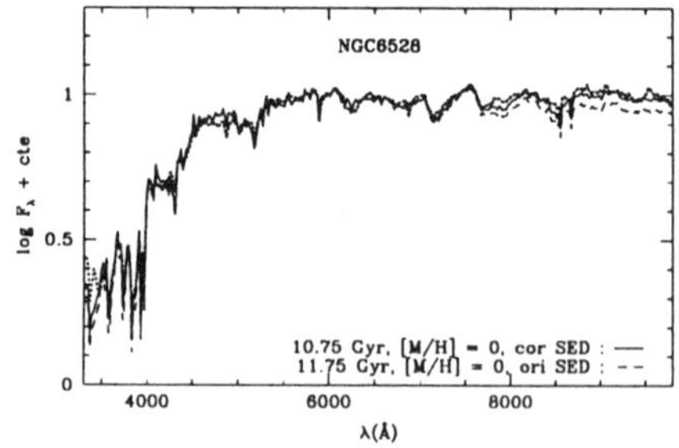

References

Bruzual, G. *et al.*, 1997,AJ 114, 1531.
Lejeune, T., 1997, PhD thesis, Univ. Louis Pasteur, Strasbourg, France.
Lejeune, T., Cuisinier, F. & Buser, R., 1997a, A&AS 125, 229 .
Lejeune, T., Cuisinier, F. & Buser, R., 1997b, A&AS, *in press*

CHEMICAL AND SPECTROPHOTOMETRIC EVOLUTION OF OPTICALLY IDENTIFIED QSO ABSORPTION SYSTEMS

U. LINDNER, U. FRITZE–V. ALVENSLEBEN, C.S. MÖLLER AND K.J. FRICKE
Universitäts–Sternwarte, Göttingen, Germany

We have compiled all published data on optically identified QSO absorption line systems (cf. Lindner, *etal.*, 1996, A&A **316**, 123). This selection with respect to gas absorption cross section supplies us with essentially unbiased galaxy samples.

Our unified chemical and spectrophotometric evolutionary synthesis code has been extended to include stellar evolutionary tracks and element yields for 5 different metallicities. This allows to follow the spectrophotometric and chemical evolution of galaxies of various Hubble types in a chemically consistent way.

Comparison of observed apparent magnitudes and colors with model calculations reveals satisfactory agreement over much of the Hubble time between our galaxy evolution models and observational data and gives information about the spectral types of the absorbing galaxies. Data in additional passbands are necessary for constraints on the cosmological parameters. Model predictions serve to facilitate future identifications.

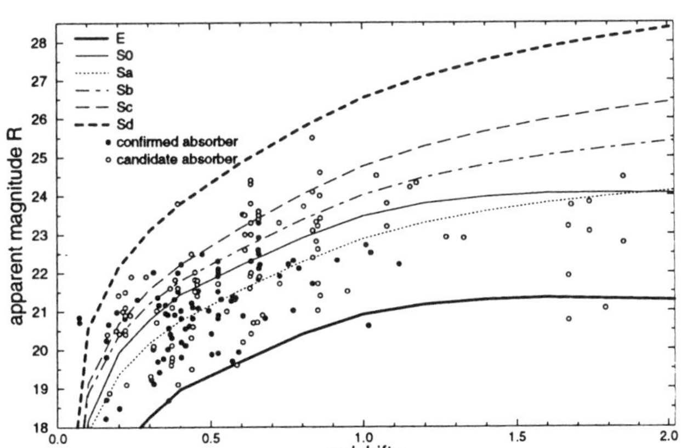

COMPARISON OF NUCLEOSYNTHESIS MODELS

O. NAKAMURA
Institute of Astronomy, University of Tokyo
2-21-1 Osawa Mitaka, Tokyo 181, Japan

Two independent sets of Type II supernova yield models, Woosley & Weaver (1995) and Nomoto et al. (1997a), are compared with each other by calculation of galactic chemical evolution of 67 stable isotopes from Oxygen through Zinc. Just the same settings of parameters such as star formation rate are used for both sets of models in calculation. Models of Nomoto et al. (1997b) are used for Type Ia supernova. Under this condition, (1) both sets can reproduce solar abundances of O, Si, Cr and Mn well. (2) Both sets show different paths in all elements at lower metallicity. Because models of Woosley & Weaver takes into account the metallicity dependence while those of Nomoto et al. offers stars of solar metallicity, it can be understood as the difference of metallicity dependence between both sets. (3) Even the models of Woosley & Weaver which takes into account the metallicity dependence cannnot completely explain observed abundances of many elements at extremely low metallicities of [Fe/H] $\simeq -4$. The figure of Cr is attached below for exsample.

reference

Woosley & Weaver,1995
ApJS, 101, 181
Nomoto et al. 1997a
astro-ph, 9706024
Nomoto et al. 1997b
astro-ph, 9706025

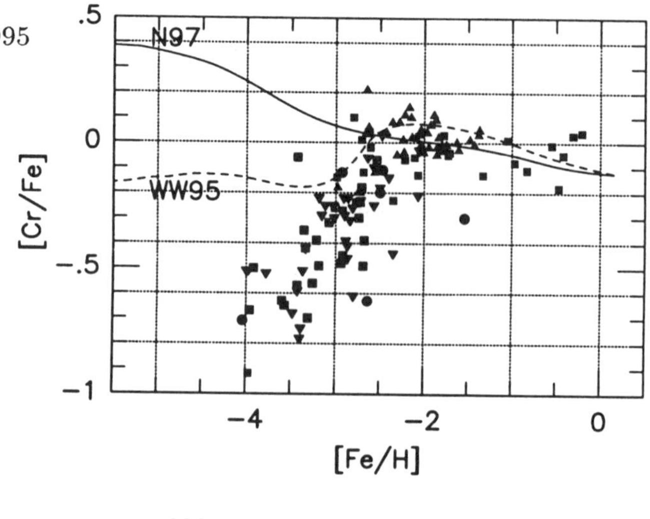

THE GALACTIC DISC AGE-METALLICITY RELATION

Y.K. NG
Padova Astronomical Observatory

AND

G. BERTELLI, G. CARRARO, L. PORTINARI
Padova Department of Astronomy
Vicolo dell'Osservatorio 5, I-35122 Padova, ITALY

Abstract. Ng & Bertelli (1998) computed new ages for the stars from the Edvardsson et al. (1993) data set. First the effects due to a change of isochrones were studied. Then the ages were derived using the Hipparcos parallaxes. The large spread in metallicity at any particular age is likely intrinsic to the disc and not due to the overlap of other galactic components. The resulting age-metallicity relation (AMR) is compared with those obtained from other methods, i.e. galactic open clusters, stellar population synthesis (star counts), and chemical evolution models. Details about the comparison of the various AMRs can be found in Carraro et al. (1997). The AMR for open clusters is corrected for the present day radial gradient: -0.07 dex kpc^{-1} (unweighted least-squares for clusters younger than 2 Gyr); within the uncertainties the average correction is independent of age. No correction for the vertical gradient was applied, since it is likely due to insufficient discrimination between age groups. The AMRs of open clusters and stars show a similar trend in the scatter and might provide clues about infalling and/or merger events.

Acknowledgements of support received from the Italian Space Agency (ASI), the Italian Ministry of University and Scientific & Technological Research (MURST), the TMR network of the EC (grant ERBFMRX-CT96-0086), and the Italian National Council of Research (CNR–GNA).

References

Carraro G., Ng, Y.K., Portinari L., 1997, MNRAS *submitted* (astro-ph/9707185)
Edvardsson B., Andersen J., Gustafsson B., et al., 1993, A&A 275, 101
Ng Y.K., Bertelli G., 1998, A&A *in press* (astro-ph/9707043)

ANALYSIS OF ELEMENTAL CARBON AND OXYGEN ABUNDANCES IN CARBON STARS – THEY HAVE TURNED OUT TO BE MORE CARBON-RICH THAN BELIEVED

K. OHNAKA
Institute of Astronomy, The University of Tokyo,
2-21-1 Osawa, Mitaka, Tokyo, 181 Japan
Present Address: Technische Universität Berlin, Institut für Astronomie und Astrophysik,
Hardenbergstr. 36, D-10623 Berlin, Germany

AND

T. TSUJI
Institute of Astronomy, The University of Tokyo

We re-analyzed elemental abundances of carbon and oxygen in three carbon stars (TX Psc, BL Ori, and V Aql) from high-resolution infrared spectra originally obtained by Lambert et al. (1986). With the interplay between the atmospheric structure and the chemical composition properly taken into account, we determined carbon and oxygen abundances self-consistently. The oxygen abundances in the three stars are solar or slightly subsolar, showing fairly good agreements with the results by Lambert et al. However, the carbon abundances derived here show marked disagreements with those derived by Lambert et al. Especially for V Aql, the carbon abundance is much higher than that they derived: the C/O ratio is as high as 2.4 in contrast to their result C/O = 1.25. These disagreements might be attributed to the difference of the model atmospheres used in the analyses or to the difference of the treatment of the excitation effect. The $^{12}C/^{13}C$ ratios derived in our previous work (Ohnaka & Tsuji 1996) and the C/O ratios derived here, including the high value in V Aql, show a very clear correlation. It strongly supports the scenario in which K and M giants evolve to carbon stars by addition of ^{12}C synthesized in the thermal pulse.

References

Lambert D.L., Gustafsson B., Eriksson K., Hinkle K.H., 1986, ApJS 62, 373
Ohnaka K., Tsuji T., 1996, A&A 310, 933

ABUNDANCES IN EVOLVED INTERMEDIATE MASS STARS

M. PERINOTTO
Dipartimento di Astronomia, Università di Firenze, Italy

AND

L. MORBIDELLI AND P. PATRIARCHI
Consiglio Nazionale delle Ricerche, CAISMI, Firenze, Italy

A sample of 285 galactic Planetary Nebulae (PNe) well studied up to 1995 has been examined to improve our knowledge on Galactocentric Chemical Gradients (GCG) and chemical enrichment of the Interstellar Medium (ISM). The sample has been subdivided following Peimbert (1978). We have examined the vertical gradients of PNe within a cylinder normal to the galactic plane with axis on the Sun and radius of 2 kpc. Excluding PNe of type IV we don't see vertical gradients. The spread of $\log(O/H)$ near the galactic plane for type II PNe is of 0.7 dex.This is larger than the maximum uncertainty of the individual determination of the abundance of oxygen, estimated to be of $\pm 50\%$ (± 0.18 dex). This supports the existence of chemical inhomogeneites in the ISM in the last $0.5 \div 5$ Gyr. The above allows to look for GCG projected on the galactic plane. For PNe of type II, GCG in usual units amount to $+0.002 \pm 0.02$, $+0.024 \pm 0.03$, -0.075 ± 0.15, -0.027 ± 0.07, -0.044 ± 0.10, -0.011 ± 0.15, -0.067 ± 0.10 for He, C, N, O, Ne, S, Ar respectively. The corresponding values from Faundez-Abans & Maciel (1987) are -0.019 ± 0.024, -0.069 ± 0.017, -0.072 ± 0.24, -0.072 ± 0.10, -0.056 ± 0.15, -0.098 ± 0.18, -0.056 ± 0.15 Note that our gradient for oxygen is quite different but within the uncertainties. We estimate the ISM chemical enrichment from PNe (Tinsley, 1978). For type I PNe we find: $He(PN)/He_{\odot} = 1.4$, $N(PN)/N_{\odot} = 4.1$. The relative mass return to the ISM from PNe comes out to be four times larger than from type II SN, with a possible uncertainty of a factor $2 \div 3$.

References

Faundez-Abans, M. and Maciel W.J. (1987) *Astrophys. Spa. Sci*,129, 353
Peimbert, M. (1978) *IAU Symp. 76: Planetary Nebulae*, ed. Y. Terzian
Tinsley, B.M. (1978) *IAU Symp. 76: Planetary Nebulae*, ed. Y. Terzian, p. 341

SPECTRAL ENERGY DISTRIBUTIONS OF $Z > 2$ GALAXIES IN THE HUBBLE DEEP FIELD

MARCIN SAWICKI & H.K.C. YEE
Department of Astronomy, University of Toronto
60 St. George Street, Toronto, Ontario M5S 3H8, Canada

Broadband spectral energy distributions ($V_{606}, I_{814}, J, H, K_s$) of 17 spectroscopically confirmed $z > 2$ Hubble Deep Field galaxies were compared with spectral synthesis models wich had been corrected for dust using the reddening law for star-forming regions. Sawicki & Yee (AJ submitted) contains the detailed description, while here we summarise the main results.

We find that the spectroscopically confirmed Lyman break galaxies in the Hubble Deep Field:

1. Are dominated by very young (typically < 0.2 Gyr) stellar populations (see Figure below). The absence of older populations implies that star formation at high z is episodic rather than continuous.
2. Have enough dust to suppress rest-frame UV flux by a factor of > 10.
3. **Have large star formation rates — typically $60h^{-2}M_{\odot}yr^{-1}$ (see Fig.).**
4. **Produce, in a typical star-forming episode, $\frac{1}{15}$–$\frac{1}{20}$ the stellar mass contained in a present-day L^* galaxy (see Fig.).**

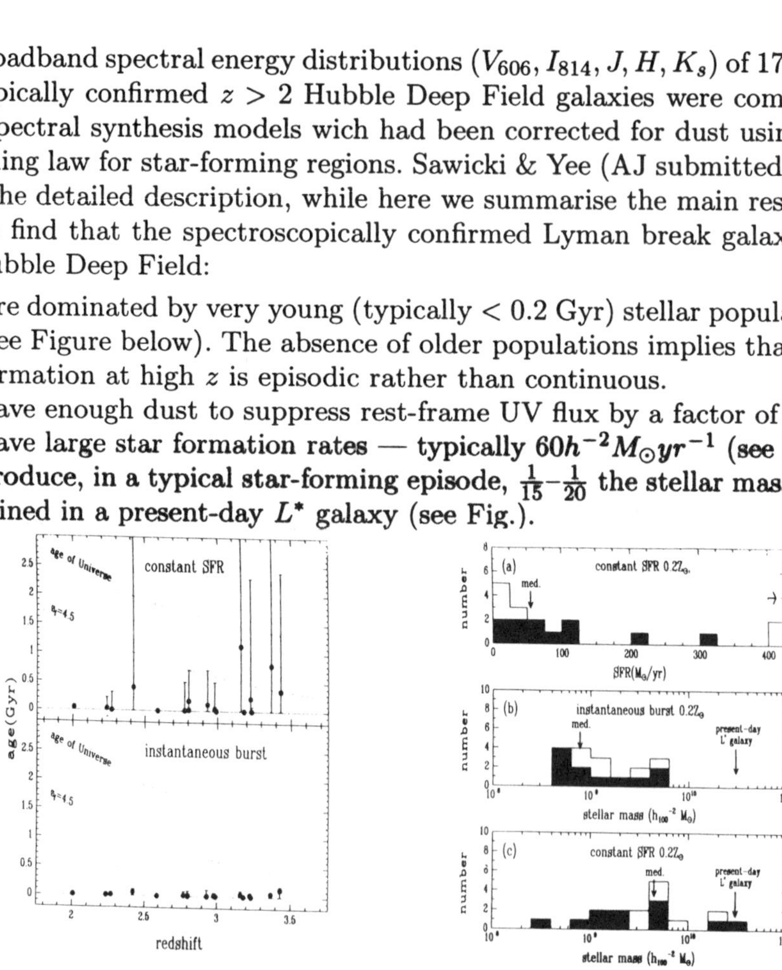

Figure 1. **Left:** Ages of dominant stellar populations from constant SFR and instantaneous burst fits. **Right:** Star formation rates and resultant masses of stellar populations.

BLUE 'ELLIPTICAL' GALAXIES IN THE HUBBLE DEEP FIELD

YASUHIRO SHIOYA AND TORU YAMADA
Astronomical Institute, Tohoku University
Aoba, Sendai 980-77, JAPAN

Abraham et al. (1996) studied the morphologically segregated galaxy count in the Hubble Deep Field (HDF) and found weak excess over the no-evolution prediction for early type (E/S0) galaxies. Dose the excess mean the mild evolution of galaxies ? To explore this problem, we studied the photometric properties of morphologically-selected early-type ('E') galaxies (ellipticals and S0s) in the Hubble Deep Field, based on the morphological catalog by van den Bergh et al. (1996). We compare the distribution of the HDF 'E' galaxies on the color - color diagram with the predicted loci expected from passive evolution for the various epoch of galaxy formation, z_f, which are derived by using the Simple Stellar Population (SSP) of GISSEL96 (Bruzual & Charlot 1996).

We found that a large fraction of these galaxies have very blue SED over the four (UBVI) band, which are never compatible with those for passively evolving old galaxies; any simple passive evolution model assuming the formation redshift which is greater than 3 cannot explain the distribution of these blue 'ellipticals' on the color - color diagram. Redshift of these blue 'E' galaxies, estimated by us with the photometric redshift method, are distributed over wide range ($z < 2$). We also attempted to fit the radial light profile of HDF 'E' galaxies using the version 2 image of HDF. More than a half of blue 'E' galaxies are well fitted by exponential law rather than de Vaucouleurs law. These results are not consistent with the interpretation of the mild evolution of early-type galaxies. The evolution of 'E' galaxies in the field seems to be more complicated than that in clusters of galaxies.

References

Abraham, R. G. et al. 1996, MNRAS, 279, L47
van den Bergh, S. et al. 1996, AJ, **112**, 359

THE SYMBIOTIC-BARIUM STARS LINK

V.V. SMITH
McDonald Observatory and Dept. of Physics, Univ. of Texas
K. CUNHA
Observatorio Nacional, Rio de janeiro, Brasil
A. JORISSEN
Institut d'Astronomie et d'Astrophysique, ULB, Belgium
AND
H.M.J. BOFFIN
Dept. of Physics and Astronomy, Univ. of Wales, U.K.

An abundance analysis of the yellow symbiotic systems AG Dra and BD-21°3873 reveal them to be metal-poor giants ([Fe/H]=-1.3) enriched in the heavy s-process elements. The heavy-element abundance distributions of those two stars are almost identical and are best reproduced by a s-process with a neutron exposure parameter of 1.2-1.3 mb^{-1} and a neutron density $\log N_{\rm n} = 8.3$. These two systems thus link the symbiotic stars to the binary barium and CH stars which are also s-process enriched. These binary systems, owe their abundance peculiarities to mass transfer from thermally-pulsing asymptotic giant branch stars, which have since evolved to become white-dwarf companions of the cool stars we now view as the chemically-peculiar primaries. We also derive the spectroscopic orbits of those two symbiotic systems and find them similar to those of barium and CH stars. Finally, because the luminosity function of low-metallicity K giants is skewed toward higher luminosities by about 2 magnitudes relative to solar-metallicity giants, we argue that lower metallicity K giants have larger mass-loss rates. It is this larger mass-loss rate that drives the symbiotic phenomena in AG Dra and BD-21°3873. A complete description of this work can be found in Smith *et al.* (1996, A&A **315**, 179 and 1997, A&A **324**, 97).

THE SPECTRAL EVOLUTION OF ELLIPTICAL GALAXIES IN FIR AND BEYOND

T.TAKAGI
Dept. of Physics, Rikkyo University, Tokyo, Japan

AND

N.ARIMOTO
Inst. of Astronomy, University of Tokyo, Japan

At the beginning, elliptical galaxies would have passed very luminous phase in the FIR, somewhat like nearby starburst galaxies and/or ultra luminous IRAS galaxies. We introduce dust into an evolutionary scheme of galactic chemical evolution by following the formalism given by Wang (1991). The evolution of dust extinction in UV-optical and re-emission at the FIR is calculated with a help of stellar population synthesis code developed by Arimoto & Yoshii (1986). Under the assumptions of spherical geometry and uniform mixing of dust and stars, we have solved the equations of the radiative transfer. Figure 1 gives spectral evolution of an elliptical galaxy with a mass of $10^{12} M_{\odot}$. As expected, the forming massive elliptical galaxies should be most luminous in the FIR wavelengths even if a gas loss due to a galactic wind is taken into account.

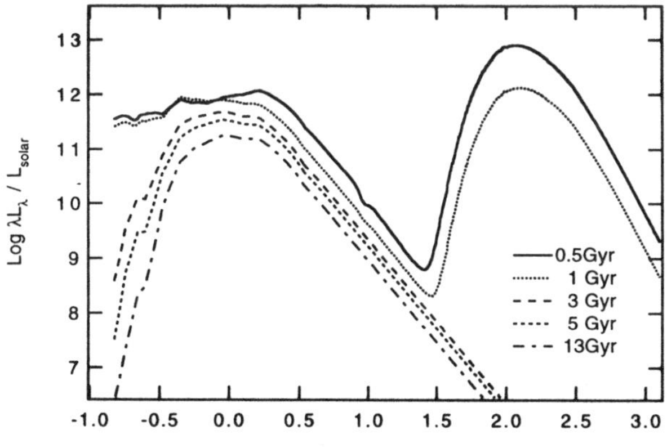

Figure 1. The spectral evolution of model elliptical galaxy with dust absorption and re-radiation. The epoch of galactic wind is assumed to be 1 Gyr.

NUCLEOSYNTHESIS IN ONEMG NOVAE: COMPARISON WITH OBSERVATIONS

S. WANAJO[1], K. NOMOTO[2], AND M. HASHIMOTO[3]
[1] *National Astronomical Observatory, Japan*
[2] *University of Tokyo, Japan*
[3] *Kyushu University, Japan*

Recent observations show that ~ 30 % of well-studied novae are classified to ONeMg novae, occurrences of outbursts with underlying ONeMg white dwarfs. There is a serious disagreement on the masses of ejected shells between observations and current theory. The estimated ejecta masses for QU Vul, V838 Her, and V1974 Cyg are $\sim 10^{-4}$–$10^{-3} M_\odot$ (Saizar et al. 1992; Shore et al. 1993; Vanlandingham et al. 1996), 10–100 times larger than theoretical estimates (Politano et al. 1995).

We have investigated the nucleosynthesis on ONeMg novae in wide range of white dwarf and envelope masses by a quasi-analytic method, using an up-to-date nuclear reaction network. Comparison of the nucleosynthetic results with observational abundance estimates enables us to constrain these two parameters. Our results suggest that at least the above three novae have obtained $\sim 10^{-4} M_\odot$ of the envelope masses, in good agreement with observational estimates. In addition, the masses of their underlying ONeMg white dwarfs may have been $\sim 1.1 M_\odot$, smaller than previously expected (see Wanajo, Hashimoto, & Nomoto 1997 for more detail).

References

Politano, M., Starrfield, S., Truran, J. W., Weiss, A., & Sparks, W. M. 1995, *ApJ*, **448**, 807

Saizar, P., Starrfield, S., Ferland, G. J., Wagner, R. M., Truran, J. W., Kenyon, S. J., Sparks, W. M., Williams, R. E., & Stryker, L. L. 1992, *ApJ*, **398**, 651

Shore, S. N., Sonneborn, G., Starrfield, S., Gonzalez-Riestra, R., & Ake, T. B. 1993, *AJ*, **106**, 2408

Vanlandingham, K. M., Starrfield, S., Wagner, R. M., Shore, S. N., & Sonneborn, G. 1996, *MNRAS*, **282**, 563

Wanajo, S., Hashimoto, & Nomoto 1997, in preparation

CONSTRAINTS ON COSMIC CHEMICAL EVOLUTION FROM THE GAMMA-RAY BACKGROUND

K. WATANABE
NASA/GSFC/USRA Code 660.1 Greenbelt, MD 20771 U.S.A.

AND

D.H. HARTMANN AND M.D. LEISING
Department of Physics and Astronomy
Clemson University Clemson, SC 29634-1911 U.S.A.

The bulk of the observed isotropic Cosmic γ-ray Background (CGB) in the MeV region measured by COMPTEL (Kappadath et al. 1996) and SMM (Watanabe et al. 1997) is believed to be due to the emission from past supernovae (e.g., The et al. 1993). γ-ray line emission from the decay $^{56}Ni \rightarrow {}^{56}Co \rightarrow {}^{56}Fe$ in Type Ia supernovae dominates due to their large production of ^{56}Ni and a higher photon escape probability in comparison to Type II supernovae. We derive the theoretical background spectrum using realistic supernova γ-ray transport models, and by integrating over the cosmic star formation history (SFH), which has recently been determined by several groups. The observed CGB places a significant constraint on the SFH. A factor three enhancement of the SFH relative to that derived from galaxy surveys would be inconsistent with the observed CGB, unless supernovae produce much lower γ-ray luminosities than current models suggest.

References

Kappadath, S.C., *et al.* ,1996, A&AS ,120,619
The, Lih-Sin *et al.* ,1993, *ApJ*,,403,32
Watanabe, K. *et al.* ,1997, *4th CGRO Symposium Proceedings, AIP*

DISK–WIND MODELS FOR BROAD ABSORPTION LINE QSO'S

T. M. YAMAMOTO
Institute of Astronomy, The University of Tokyo, Mitaka, Japan,
e-mail: yamamoto@mtk.ioa.s.u-tokyo.ac.jp

Broad Absorption Line (BAL) spectra of QSOs show a wealth of metal absorption lines which can help to identify enrichment mechanisms in the early universe. BAL QSOs are believed to be viewed from the side through a gas streaming outward from an accretion disk around the central source. Models for BAL QSOs based on Monte Carlo simulations are introduced, in which the gas is confined to spherical or disk geometry.

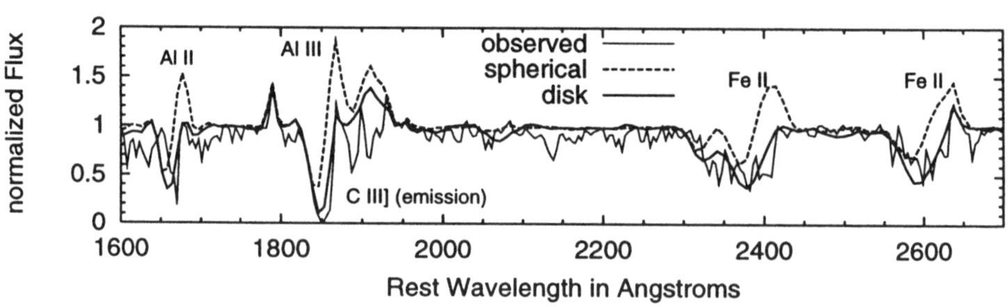

The figure shows a comparison of two synthetic spectra with an observed spectrum of the BAL QSO Q0059–2735 (thin line). The spherical model (dashed line) has too weak absorption lines and strong emission on the red wing. The emission on the red side is caused by the same phenomenon as in P–Cygni objects where we see the backward expanding shell behind the central source. In the disk model (thick line) this expanding gas can not be seen because we look from the side into the disk plane. The absorption, on the other hand, is deeper in the disk model, because each line scattering event most likely scatters the photon away from the disk plane. From this it is concluded that the disk model better reproduces the observed spectrum when compared with the spherical model spectrum.

This investigation of the outflowing gas near to the central region of QSOs shows that the appropriate geometrical structure of the absorbing gas is important for the analysis of broad absorption line regions.

SUPERNOVA OBSERVING WEB (SNOW) PROJECT WITH THE PUBLIC ASTRONOMICAL OBSERVATORIES IN JAPAN

HITOSHI YAMAOKA
Department of Physics, Faculty of Science, Kyushu University 4-2-1 Ropponmatsu, Chuo-ku, Fukuoka 810, Japan
MAMORU DOI, TOSHIKAZU SHIGEYAMA, NAOKI YASUDA
Department of Astronomy, School of Science, University of Tokyo, 2-11-6, Yayoi, Bunkyo-ku, Tokyo 113, Japan
AND
MASARU WATANABE
Astronomical Data Analysis Center, National Astronomical Observatory Japan, 2-21-1, Osawa, Mitaka, Tokyo 181, Japan

We are promoting a supernova(SN) detecting project (named SNOW) with medium size (60cm – 105cm) telescopes belonging to the public astronomical observatories in Japan. Main observations were started on January 1997 with the cooperation of seven observatories, and will be continued some years with more participants. One of our purposes is to determine the SN rate especially in early type galaxies, which plays an important role in the chemical and dynamical evolution of them and surrounding hot gases. Each telescope, with the R-band filter and chilled CCD camera, observes several fields allotted respectively for it. Fields are selected to include the dense cores of nearby ($z < 0.1$) clusters of galaxies, then each field contains about 10 – 20 regular E/S0 galaxies. In order not to miss the SNe emerging in the fields, every fields are checked within two weeks after the latest observation. The flatfielding and the detection of new (or brighter) objects are automated. To inform the monitoring status for compensating each other for lack of observation, a WWW page is organized (`http://www.netwave.or.jp/~yfujita/snow.html`, written in Japanese). It is also used for the publicity of the project. Unfortunately, we have not detected any new SNe in our field till October 1997. We expect that we will be able to determine the SN rate within the 35% error after two years spell, which will be reduced with a longer run or more entrants.

ELEMENT ABUNDANCES IN METAL-POOR STARS

G. ZHAO
Beijing Astronomical Observatory, Chinese Academy of Sciences, Beijing 100080, China

AND

Y.Q. CHEN
Department of Astronomy, Beijing Normal University Beijing 100875, China

We analyzed the observational material of high resolution spectra for a sample of 8 metal-poor stars. The data were collected with the Coude Echelle Spectrograph attached to the 2.16m telescope at Beijing Astronomical Observatory (Xinglong, China). The atmospheric parameters of these program stars have been determined. The effective temperature was derived from the color index or literatures. The surface gravity was determined by forcing the lines from two ionization stages of the same element to indicate consistent abundances. The micro-turbulent velocity was determined by forcing all the FeI lines with equivalent width between 0.1-0.8 nm to give the same abundance. We adopt a classical method of analysis, the abundance being deduced from each line by forcing the computed EW to agree with the observed one. The model atmosphere is interpolated in the grid of Magain, which was computed with a version of the Gustafsson's program. Overall abundance patterns of 19 important elements, such as light elements, iron peak and heavy elements, are deduced on the basis of the data reduction and line analysis. The light and iron peak elements are generally agree with the analysis of Zhao and Magain(1990). But aluminium abundance in HD 19445 is very low, [Al/Fe] $= -1.52$ which is 0.77 dex lower than that of Magain (1989). The heavy elements behavior agree with most investigations.

References

Magain, P. (1989) *A&A*, **209**, 211
Zhao, G., Magain, P. (1990) *A&A*, **238**, 242